**SPARSHOLT COLLEGE HAMPSHIRE**
**Library and Information Centre**

~~ks~ are to be returned on or before
~~the~~ last date ~~b~~

returned on or before
~~dat~~e below

# —The Manual of—
# *LIZARDS*
# *& SNAKES*

# The Manual of
# *LIZARDS*
# *& SNAKES*

## MARC STANISZEWSKI

## Tetra✺Press

No. 16048

Left: *A light form of the distinctive Trans-Pecos ratsnake,* Elaphe subocularis.

# A Salamander Book

© 1990 Salamander Books Ltd.,
Published by Tetra Press,
201 Tabor Road,
Morris Plains, NJ 07950.

ISBN 3-89356-040-8

All correspondence concerning the content of this volume should be
addressed to Tetra Press.

### The author
Marc Staniszewski has been fascinated by herptiles since early
childhood and has been studying them for the last 17 years.
Encouraged by his father, he has kept and photographed many of the
species described in this book. He has travelled throughout Europe,
studying indigenous species and is a regular contributor to specialist
magazines in the UK. He lives and works in Worcestershire, England.

### Credits
Edited and designed by Ideas into Print,
Vera Rogers and Stuart Watkinson.
Typeset in Palatino on a Macintosh IICX. Bromides: Scarbutts.
Paste-up: Steve Linehan. Film conversion: SX Composing Ltd.
Colour reproductions: Scantrans Pte. Ltd. Printed in Singapore.

# Contents

# Part one:
# Lizards

The lizards that inhabit the world today are the result of 300 million years of evolution. They represent the most successful and adaptable of all reptiles, having survived many climatic and environmental changes since their ancestors dominated the earth for 100 million years. Indeed, in terms of distribution, appearance, habitat and habits, lizards are more diverse than any other group within the class Reptilia.

Nevertheless, the explosion in the human population has put extreme pressure on wild populations of both reptiles and amphibians (collectively known as herptiles). Afforestation and the introduction of domestic animals - cats, for example, or vermin, such as rats - have combined to threaten their traditional habitats. Although some species have been able to colonize man's apparently hostile 'concrete jungles', other populations have declined so dramatically that extinction can be only a matter of time. Particularly rare are the day geckos, *Phelsuma* spp., from some of the smaller Indian Ocean islands, the giant skink, *Macrodiscus cocteaui* from the Cape Verde Islands in the Atlantic, and the Solomon Islands giant skink, *Corucia zebrata*. Many countries now ban the export of lizards.

On the positive side, people are taking more interest in herptiles and putting aside their previous misconceptions. For example, most lizards are neither poisonous (only two species are venomous) nor dangerous if handled correctly. On the contrary, in the wild they play a beneficial role, controlling harmful insect populations and thus reducing the incidence of disease and crop destruction. They are an important part of the food chain, forming the main diet of many higher vertebrates.

As we learn more about lizards, the techniques of captive maintenance and breeding are becoming simplified, and improved equipment is appearing on the market. In the practical chapters of this book, we examine all aspects of caring for these fascinating creatures in captivity, from buying, handling and housing to feeding and breeding them. Because of the threat to lizard populations and the ban on certain exports, as well as the increased demand for captive stock, there has never been a more important time to care for and breed these creatures. Fortunately, most lizards are easy to maintain and breed, but some species may tax the capabilities of the most experienced herpetologist. However, with such a wide range of species to choose from there is less temptation to keep the 'difficult' species. The species section features a survey of lizards that adapt well to life in a vivarium.

# WHAT ARE LIZARDS?

The suborder Sauria is the largest group within the order Squamata, the only other suborder being the snakes, Serpentes. Today, approximately 3,000 known species are found throughout the world, from northern Sweden to the southern tip of South America.

Lizards vary greatly in size; *Sphaerodactylus parthenopion* from the West Indies rarely attains a total length of 3.75cm(1.5in), while the famous Komodo dragon, *Varanus komodensis*, can grow to 3m(10ft) long and weighs over 136kg(300lb). However, this is not the longest lizard in the world; that honour belongs to the highly aggressive Salvator monitor, *Varanus salvator*, an Indonesian species that can grow to more than 4.5m(15ft).

Lizards are not renowned for being long-lived, but a few species do reach a respectable age. The slow-worm, *Anguis fragilis*, a common European legless lizard, can live up to 50 years. The only other lizard to come anywhere near this age is, once again, the Komodo dragon, which lives for about 25 years in the wild.

Speed plays an important part in the fight for survival, particularly among the smaller lizards. The six-lined racerunner, *Cnemidophorus sexlineatus*, from central and eastern USA, can attain short bursts of 29kph(18mph), and the collared lizard, *Crotaphytus collaris*, from central and southern USA, can run on its hind legs at a speed of 26kph(16mph).

## Lizard anatomy

The skeleton of a lizard is a fairly simple construction, with a well-developed shoulder and pelvic girdle, even in those species without external limbs. The thickness and shape of the skeletal bones varies with the habits of the lizards; for example, the burrowers have a thick strong skull for their excavation activities, whereas arboreal

## Lizard anatomy

*In some lizards, the long tail can be used to store fat. Other species voluntarily lose their tails in order to escape predators.*

*A lizard's watertight scaly skin provides protection in the most hostile environments.*

*The ear opening, or external tympanum, is visible and can detect most frequencies.*

*The eyes are well developed and some species have excellent nocturnal vision.*

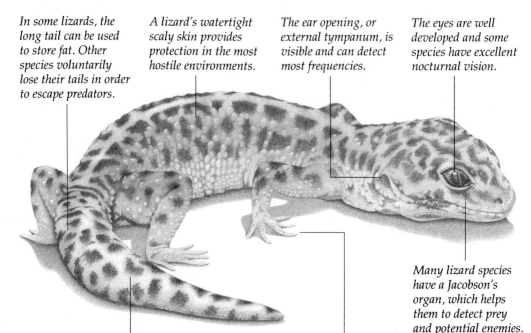

*Many lizard species have a Jacobson's organ, which helps them to detect prey and potential enemies.*

*The leopard gecko's markings fade with age. Other factors affect lizard coloration, including temperature variation and breeding behaviour.*

*Different lizard species have developed toes, pads, claws or stumps, depending on the terrain in their natural habitat and their lifestyle.*

species are equipped with strong tail and limb bones, while the rest of their skeleton is lightweight. The associated muscles show the same variation from species to species.

**The ribs** In many species, the ribs are movable, enabling the lizard to extend its body outwards. By flattening itself in this way, the lizard can absorb more of the sun's warmth or, alternatively, hide from the midday sun (and predators) by crawling into tiny crevices or beneath rocks. The flying dragon of the genus *Draco* has highly mobile and flexible ribs that enable it to glide in any direction by arching or flattening single ribs.

**The toes** Many lizards have modified pads under the toes. Geckos, for example, have a number of overlapping flaps, or lamellae, which are clothed with microscopic, backward-pointing hooks. The hooks are so small that they can grip onto the tiniest irregularity, and their combined effect is more than enough to hold the gecko on a ceiling or even a vertical window pane.

Above: *Some lizards, such as the tree agama,* Acanthogaster lepidogaster, *are adorned with horns and frills. Their main function is to deter predators or attract the opposite sex.*

Right: *Like most lizards, the skink,* Mochlus fernandi, *enjoys sun basking.*

Below: *This close-up clearly shows the specialized toe pads found in many geckos.*

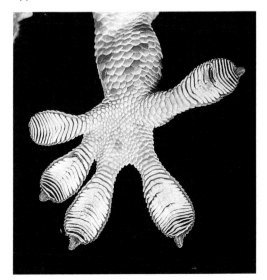

**The tail** Lizards are perhaps best known for 'losing' their tails. This voluntary amputation, known as autotomy, is a survival mechanism adopted by the majority of smaller species, particularly the lacertids and skinks. Between each vertebra is an extremely weak and brittle area - the intravertebral plane. If the tail is seized, it separates from the body at one of these points, automatically severing the associated muscles and blood vessels. The muscle in the severed tail contracts, causing it to wriggle and confuse the predator, all of which gives the lizard an opportunity to escape. The tail regenerates slowly afterwards, consisting not of bone but of a rod of cartilage, but is never the same colour or length as the original one.

**The teeth** The majority of carnivorous lizards have a single row of short, pointed teeth fused directly onto the jawbone. Specialization in the teeth is less common than in mammals. Planteaters, such as certain iguanas, have flat grinding teeth and only the omnivorous agamas and teiids have incisors, canines and molars.

**The skin** Depending on its natural habitat, a lizard's skin may be either smooth or rough, but it is always dry. Burrowers have small smooth scales that cause no resistance, whereas land- or tree-dwellers have thick, rough or horny scales to protect them from the natural conditions. Similarly, the thickness of skin varies with geographical

## Lizard families and their distribution

| Family | Number of species | Common name and characteristics |
|---|---|---|
| Agamidae | 300 | Agamas. Huge and diverse family found in the drier areas of Eurasia, Africa and Australasia. |
| Amphisbaenidae | 130 | Ringed lizards. Subterranean legless lizards from Africa, Asia and Central to South America. |
| Anelytropsidae | 1 | Mexican blind lizard. Rare burrower. *Can* see. |
| Anguidae | 70 | Slow worms; Alligator lizards. Diverse family, but share similarities in tongue/bone structure. |
| Anniellidae | 2 | Californian legless lizards. Livebearing burrowers. |
| Bipedidae | 3 | Two-legged worm lizards. Burrowers with just two strong forelimbs. From California and Mexico. |
| Chamaeleonidae | 80 | Chameleons. Predominantly African family, mainly arboreal and masters of mimicry. |
| Cordylidae | 33 | Plated lizards; Girdletail lizards. Robust rock dwellers from Africa. |
| Dibamidae | 3 | Wormlike creatures found in the Malay Peninsula. |
| Feyliniidae | 4 | Little-known skinklike lizards from Africa. May be merged into the subfamily Scincidae. |
| Gekkonidae | 550 | Geckos. The only truly vocal lizard and an expert climber. Found throughout warm regions. |
| Heliodermatidae | 2 | Bearded lizards. The only two venomous lizard species. From western USA and Mexico. |

distribution; lizards from desert regions have several thick epidermal layers to reduce moisture loss, while species from temperate regions have thinner skin to allow the sun's warmth to penetrate more quickly.

**The eyes** Apart from burrowers and certain 'blind' lizards, eyesight is well developed in lizards, but it varies in acuity. Chameleons and geckos have the most sensitive vision. Unlike snakes, lizards generally have movable eyelids, but in some of the burrowers the eyelid is permanently closed, forming a protective transparent covering.

**Jacobson's organ** Terrestrial and arboreal lizards rely mainly on eyesight for hunting, but many species have another highly developed sensory organ, more commonly

Above: *The beautiful collared lizard,* Crotaphytus collaris, *which inhabits desert* *areas in southwest USA, can run quickly on its hind legs if alarmed or excited.*

| Family | Number of species | Common name and characteristics |
|---|---|---|
| Iguanidae | 670 | Iguanas. A huge, diverse family, consisting of five subfamilies. Found throughout the Americas and also in Madagascar and Fiji. |
| Lacertidae | 200 | Typical lizards. Able to shed tail. From Europe, Africa and Asia. |
| Lanthanotidae | 1 | Earless monitor. Anatomically similar to snakes. Found only in Sarawak, Borneo. |
| Pygopodidae | 13 | Scaly foot. Snakelike lizard from Australia, anatomically related to the gecko. |
| Scincidae | 700 | Skinks. Huge, diverse family found in the warmer regions of every continent. |
| Teiidae | 200 | Racerunners; Tegus. Found throughout the Americas and similar to the family Lacertidae. |
| Trogonophidae | 8 | Sharptailed worm lizards. Legless burrowers from southern Europe, Africa, southeast USA, Central and South America and the West Indies. |
| Varanidae | 31 | Monitors. This family includes the largest lizards. All are muscular, with elongated bodies and snakelike tongues. |
| Xantusidae | 12 | Night lizards. Found in the American deserts and the West Indies. |
| Xenosauridae | 4 | Crested lizards. Little-known lizards from China and Central America. |

associated with the snakes. This is the Jacobson's organ, which consists of two cavities in the roof of the mouth, lined with an extremely sensitive membrane. As the tongue flicks out, it picks up microscopic odorous particles from the air and ground, and from animate and inanimate objects. When the lizard withdraws the tongue into the mouth, it inserts the tip into the cavities, where the sensitive membrane registers the chemical 'taste' of these particles. (In lizards with forked tongues, the tips are inserted into both cavities.) Slow-moving lizards, particularly monitors, have a well developed Jacobson's organ. However, members of the chameleon and agamid families have developed differently; they use their mucus-covered tongues for catching prey.

### Regulating body temperature
In common with other herptiles, lizards are ectotherms, meaning that their body temperature varies according to the temperature of their environment. Nearly all species rely on the heat of the sun to provide their energy and keep them active, so the further north or south from the equator you travel, so the number of species diminishes, until only those species able to endure long periods of inactivity by hibernating during the colder months are found at the extremes. Generally speaking, a lizard's critical body temperature lies within the range of 25-35$^0$C(77-95$^0$F). In some temperate species it may be as low as 16-27$^0$C(61-80$^0$F), whereas in desert species it is between 23 and 38$^0$C(73-100$^0$F). Most lizards become inactive when

the temperatures drop to 11$^0$C(52$^0$F) and will die if it falls below 1 or 2$^0$C(34-36$^0$F).

In desert regions, there is often a drop in temperature at night, severe enough to kill an ectotherm. In fact, a lizard can survive by seeking the protection of rocks and similar dense objects with a large surface area. These retain the heat and may be as much as 10-15$^0$C(18-27$^0$F) warmer than the surroundings - sufficient to maintain a lizard's critical body temperature. This interaction is known as thigmothermy, from 'thigmo' meaning to touch and 'therm', meaning warmth.

### Hibernation
The ability to hibernate is very important to species from areas that experience severe wintry conditions and, therefore, a scarcity of food. Here, lizards - often together with other reptiles, such as snakes - retreat under trees or into frost-free, disused burrows, known as hibernaculums. Their metabolic rate reduces significantly and they rely on existing body fat for energy. Should the temperature rise during the hibernation period, the lizard's metabolism starts to increase; in such conditions some lizards become active and often bask in the weak midwinter sun.

### Aestivation
In subtropical regions, such as the southern Mediterranean and southern USA, severe heat and drought force the lizards deep into the shelter of cool burrows. If their body temperature rises to 43$^0$C(110$^0$F) they invariably die. In a state of aestivation the metabolic rate is the same as in hibernating

Left: Lacerta lepida, *the eyed lizard, and other European lacertids are diurnal, sun-loving lizards. A drop in temperature stimulates breeding.*

Right: *The livebearing slow-worm,* Anguis fragilis, *produces between 6 and 24 coppery coloured young. This striking coloration fades as the juveniles mature.*

lizards, but many species awake from this semi-torpid state to bask in the cooler dawn or late evening sun. In captivity, hibernation or aestivation act as a stimulus for breeding in certain lizards (see also page 49).

## Reproduction

Reproduction in lizards follows a basic pattern, but often requires a stimulus before it can begin. Stimuli include increased activity after hibernation or aestivation, pheromones ('scent hormones') produced by the female after skin shedding, an increase in the food supply or climatic changes. In some species, there is no special courtship display, but in species where males usually heavily outnumber females, the males will compete against each other for the right to mate with - or gain a territory containing - a female. Lizards frequently fight and indulge in furious displays of body movement, such as tail waving, foot stomping or even jumping up and down, and can show dazzling breeding colours. The dominant male then begins a courtship ritual to attract the female's attention. This may involve head bobbing or softly biting her, sometimes causing minor injury. Copulation takes place within a few seconds, the male often grabbing the female by the base of the tail with his jaws and quickly inserting his reproductive organ, the hemipenis, into her cloaca and releasing small sacs of sperm.

If the lizard is egg laying (oviparous), the next stage is a relatively short gestation period before the eggs are deposited in a warm, moist place. Although the eggs have a

Above: *A green lizard,* Lacerta viridis, *hatches out from its* egg. *These early days are the most critical time in its life.*

tough, parchmentlike shell, this offers only limited protection against water loss, so a damp position guards against the risk of eggs becoming dehydrated. The incubation period varies considerably according to species and ambient temperatures and hatching takes place after 30 to 90 days.

Live-bearing (viviparous) lizards have a longer gestation period before they deposit translucent membranes containing their young. These push their way out after absorbing the remainder of the yolk-sac and begin their perilous lives without parental protection. In both types of reproduction, the hatchlings are usually miniature replicas of their parents, although juvenile coloration may be brighter and more well-defined at this early stage.

# BUYING AND HANDLING

The steady and inevitable decline of lizards through loss of habitat, over-collecting for the pet trade or persecution is quite alarming. Fortunately, governments in Europe, North America and Australia have introduced strict legislation to control, or even ban altogether, trading in all wild animals, including lizards.

Such protection includes the International Conventions and National Legislation (Washington 1973), which lists nearly extinct, endangered and rare animals and plants, and restricts or prohibits their trade. To begin with, few lizards were listed, but research and tighter control has increased the number significantly. The Berne Convention of 1979 set out to protect European herptiles and, with many countries introducing local legislation, has succeeded in effectively stopping the trade of all species. Indeed, nearly all European species on sale today are captive-bred.

Nevertheless, wild-collected specimens are still available, particularly species from tropical and developing countries. In some cases, wild populations are so common that it is cheaper for dealers to supply imported specimens than to obtain captive-bred stock.

In the interests of conservation, it is clearly desirable to buy specimens bred in captivity. However, there are other good reasons for doing so. Although wild-caught specimens may look healthy, they often have internal parasites or infections, and they may not adapt to life in captivity, refusing the food substituted for their natural diet, for example. (However, it is true that some species, particularly geckos and small lacertids, do adapt to captivity almost immediately.) Captive-bred lizards, on the other hand, are conditioned from birth to live in an 'artificial' environment and are likely to be relatively or completely free of disease.

## Buying lizards
Today, trade in herptiles is at an all-time high, especially now that we know more about their breeding requirements. In the last 10 years the number of dealers - general pet stores, specialist and hobbyist dealers and mail order companies - has increased fourfold in western Europe and the USA and deciding where to buy can be difficult.

Although the local pet shop is unlikely to specialize in reptiles, it is still a favourite source of supply. The species offered for sale are generally easy to maintain, healthy and well looked after. The assistants are usually helpful, even if they are not able to offer expert advice.

The specialist dealer usually has a good, if not huge, selection of different species, together with a wide range of equipment. A good dealer will be able to offer sound advice on difficult species, all aspects of their captive care, and will be able to sex lizards. Most importantly, you will be able to see the specimens for yourself and choose accordingly. Wherever you buy your lizards, make sure that they are kept in good, clean conditions and are not overcrowded. Responsible, knowledgeable dealers and pet shops are far more likely to sell strong and healthy animals.

Mail order is an increasingly popular way of obtaining both common and rarer lizard species. It is often relatively inexpensive, but remember that delivery charges and, perhaps, insurance will raise the price. Both professional dealers and hobbyists with relatively small stocks may sell by mail order. The main disadvantage of buying your lizard this way is that you will not be able to view it beforehand, so make sure there is a clause in the advertisement stating that you can return the lizard for a refund or replacement if you are not satisfied with it. Specialist magazines often carry advertisements for mail order companies selling lizards and may offer a degree of protection should you encounter difficulties after buying through such advertisements.

Joining a herpetological society will give you access to an excellent source of stock. Most societies produce a newsletter which, as well as providing up-to-date information on herptiles and their care, usually has a 'for sale' section, often listing a good selection of captive-bred and other healthy lizards. At society gatherings, you will meet other enthusiasts who may have lizards to sell.

Above and left: *Two commonly kept lizards that differ in size, appearance, habits and habitat.* Iguana iguana *grows up to 120cm(48in) and is found in warm, humid forests around equatorial America, whereas the 13cm(5in)* Lacerta vivipara *is an inhabitant of northern Europe, living in dry open meadows.*

## Choosing a suitable species

Before acquiring a lizard, it is a good idea to find out as much about it as possible. Given the scarcity and high cost of some species, you want to be sure that you are buying a healthy animal and can provide it with the best possible living conditions. Your local library, specialist societies and magazines are all useful sources of information.

Some species are best avoided by novice keepers. They may be expensive, short-lived, highly aggressive or difficult to cater for. As you gain experience, you may decide to attempt some of these species but, even then, certain ones are only suitable for zoos or specialized collections.

These are some of the factors to bear in mind when buying stock.

**Size** If you are buying a juvenile or subadult specimen, find out how large it will be when fully grown. Can you provide accommodation of a sufficient size?

Do not overlook the risk of being attacked by a large lizard, perhaps over 90cm(36in) long or longer, especially if there are small children about. Although tegus, *Tupinambis teguixin*, red tegus, *T. rufescens*, Caiman lizards, *Dracaena guianensis* and most monitors are usually acquired as small, harmless-looking specimens, they will eventually grow into potentially aggressive animals, capable of inflicting serious injury with their strong jaws or tail. This warning applies mainly to the larger lizard species, but even some of the smaller lizards, such as the Tokay gecko, *Gekko gecko* have an aggressive nature. When you first handle the larger species, you can take the precaution of wearing leather gloves, but take great care not to injure the lizard by exerting too much pressure - remember that wearing gloves reduces your sensitivity. If possible, always handle lizards with bare hands.

**Specialized feeders** Live food items may be difficult to obtain during the colder or drier seasons of the year, but most species will adjust to an alternative diet. However, in certain lizards this is simply not possible; their diet is so specialized that it is either difficult to obtain or not available in sufficient quantities.

**Habits** Consider the lizard's social habits. Is it solitary and secretive or a gregarious species that prefers to live in a large group? Does the species form distinct territories or does one animal dominate a group? Is it nocturnal or diurnal? Does it enjoy long periods of activity all year round or does it need to hibernate or aestivate?

**Habitat** What sort of terrain and climatic conditions prevail in the lizard's natural habitat? Will it be possible to recreate these conditions adequately in captivity? Lizards occur in a wide variety of habitats and in many cases their requirements can be met.

Below: *Small lizards, such as this Sumatran skink,* Mabuya, *should be held gently with a bare hand to avoid causing internal bruising. When handling larger, more temperamental lizards, you may need to protect your hands with leather gloves.*

### Short-lived species
Some members of the chameleon family are extremely short-lived, surviving no longer than 12-16 months. This may be a reflection of their natural lifespan or because they require specific micro-climates within their captive environment. As they are expensive to acquire and often difficult to breed, they are not recommended.

### Choosing healthy stock
A healthy lizard should appear alert and eager to avoid capture. It should be well-fed but not obese, and its skin, mouth, eyes and digits should be in good condition. Look for signs of excessive scarring, loss of a limb or evidence of parasitic infection (see also *Health care,* page 56).

If you are buying more than one specimen, choose lizards of roughly equal size, in order to avoid potential bullying. Juvenile or subadult lizards adapt far better to a captive environment, but may be difficult to sex. Ideally, sex ratios should be equal for most species, but where individuals are known to fight, aim to keep a single male with a number of females.

Always quarantine a newly obtained lizard, i.e. keep it isolated from other lizards for six to ten weeks before introducing it to its new home. This will give it time to recover from the stress of moving into a new environment and gives you a chance to ensure that it is healthy and feeding well. Simply place the lizard in a warm cage with appropriate lighting, some newspaper on the base, a few rocks and fresh drinking water.

### Handling lizards
Generally speaking, you should handle lizards as little as possible, as they are easily stressed and damaged. There will be times when it is unavoidable, such as when you clean out the vivarium, measure or weigh the lizard for statistical analysis, treat it for disease or clip its claws.

It is particularly important to avoid handling lizards that are sloughing their skins - they are likely to be rather irritable and may bite or struggle. Handle heavily gravid (pregnant) females with great care, especially egg-layers. It is safer to scoop them up into a box, guiding them carefully with your free hand than to grasp them suddenly.

Above: *Many of the* *muralis, shown here,*
*skinks, teiids and* *lose their tails easily.*
*lacertids, including the* *Eventually, a new but*
*wall lizard,* Podarcis *smaller tail will grow.*

Disturbing a torpid lizard can prove fatal; always warm it thoroughly to induce an active state before handling it.

When confronted by a potential enemy (including man) a lizard's first instinct is to escape or hide. If it is regularly handled, it gradually loses this instinct and becomes tame, but this can have adverse consequences. It may become lazy, lacking its usual alertness and the quick responses associated with wild specimens. For example, the lizard will rarely try and hunt for food, waiting instead until its keeper brings food to its jaws. Breeding behaviour may also be affected; it will not attempt to mark out a defined territory or respond to the opposite sex - it may, in fact, become too lazy to breed. Some lizards - the popular blue-tongued skink, *Tiliqua*, for example - is naturally docile and appear to enjoy the attention of humans, but even in these cases, handling is best kept to a minimum.

When caught, many species will defaecate profusely, so it is best to wear protective clothing and have plenty of absorbent tissue to hand. Place newspaper around the vivarium to avoid soiling the floor.

**Handling legless lizards**
Gently place one hand round the neck of these snakelike lizards and support the rest of the body in the other. Never grab anguids or legless skinks by the tail, which is very fragile and will break at the slightest touch. Some of the larger legless species, such as the muscular glass lizards, *Ophisaurus* sp., require more restraint, as they can inflict a rather painful bite. It is a good idea to wear gloves when handling large specimens.

**Handling small to medium-sized lizards**
The majority of lizards, including lacertids, agamids and skinks, fit into this category, measuring up to 60cm(24in) long. The smaller varieties, in particular, are extremely agile and may be difficult to catch. Again, take care not to grab the tail and bear in mind that the skeleton is fragile and easily crushed.

Once safely captured, the best way to handle smaller lizards is to place the thumb and forefinger of the dominant hand around the neck, applying sufficient pressure to restrain but not suffocate it. Lay your palm along the lizard's back and the remaining three fingers underneath the belly and hind limbs. Use your other hand to stabilize the tail by gently holding the base.

**Handling larger lizards**
Handling a lizard over 60cm(24in) long can be difficult. You must not only restrain it, but avoid being scratched, bitten or whipped by the tail. Docile species, such as blue-tongue skinks, present no real problem, but even they are endowed with powerful jaws and sharp teeth. You may need help when handling the green iguana, tegu and various monitors and you should certainly wear leather gloves and strong clothing that cannot rip or snag.

Hold the head with your non-dominant hand and the tail with the other or, if the lizard is particularly large and powerful, grasp the tail between your legs, leaving your hands free to support the abdomen and head. Another person can then examine the specimen or guide it into a container so that you can clean the vivarium.

# HOUSING LIZARDS

Lizards are normally housed in vivariums. These range from a simple glass jar to a large and elaborate set-up and your choice will depend on the size of the species you intend to keep. The aim is to provide an environment in which the lizard can live and behave naturally and does not become stressed. In this chapter, we examine some of the options available.

### Small containers
Many small lizards, including geckos and lacertids, are easier to maintain and study if they are housed in a small container, such as a margarine tub, glass jar or plant propagator. Such containers are ideal for newly hatched or juvenile lizards of many species, because each one can be housed separately, given individual attention and spared the stress of competing for food.

Whatever the container, good ventilation is essential. If you opt for a plastic tub or plant propagator, simply cut out a 7.5cm(3in) diameter hole and then cover it with a slightly larger piece of aluminium mesh (obtainable from a car accessory supplier). Fix the mesh in place by heating the edge

Below: *Before buying or making a vivarium, find out as much as you can about the* habits of your lizard. Eumeces schneideri, for example, is a ground-dwelling skink.

over a flame until it glows and then placing it over the cut hole, where it should fuse to the plastic. If you use a glass jar, remove the lid and replace it with muslin firmly secured with a rubber band. Remember that lizards are expert at escaping, so ensure that all possible points of escape are blocked.

Line the base of the container with a mixture of soft tissue paper and newspaper and supply extra shredded newspaper in which the lizard can hide. A small water container is essential.

### Display vivariums
In Europe and North America it is possible to choose from a wide range of vivariums especially designed for herptiles. Some are made entirely of glass, while others have a glass front and solid sides, but all come in a variety of shapes and sizes. All-glass vivariums (or aquariums) are suitable for burrowing species, or those that require very humid conditions, and they can be impressively furnished with live plants. Lids are available in various sizes, equipped with a sliding door, ventilation grill and a hole for light fittings. All lids must be escape-proof.

Many lizards, including small lacertids and skinks, are nervous creatures, often stressed by the 'exposure' of an all-glass aquarium. They will feel more secure in a glass-fronted vivarium with solid side and back panels, which you can buy ready made. The most elaborate models are equipped with time-controlled lighting, thermostatically controlled heaters and humidifiers. These vivariums are very efficient but an expensive investment, so you may wish to design and assemble your own.

### Building a vivarium
Whether you opt for a simple or an elaborate vivarium, remember that it should be easy to clean, escape-proof and designed to meet the needs of the species it is to house. Suitable materials include metal and plastic, but wood is the most adaptable and easy to handle. Chipboard or plywood (about 10mm/0.4in thick) are equally suitable, but both will require waterproofing. One method is to

## Building a vivarium

Buy and saw the wood carefully to the exact measurements. Drill all the holes required for screws, cables and ventilation grills.

Fit all the side panels together and, if the timber is not already treated, apply a coat of waterproof paint. Leave until dry.

Cut strips of double plastic or wooden runners, and attach to inside edge of roof and base panels with a strong adhesive.

Paint roof panel with waterproof paint. Set aside roof and glass doors until lamps and holders are installed (see page 27).

  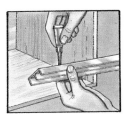 

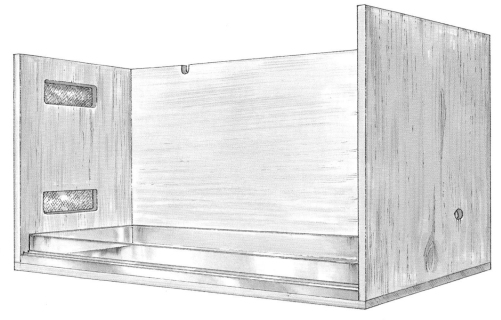

 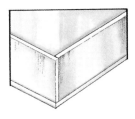  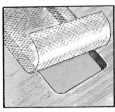

Seal all the joints and corners of the vivarium with waterproof aquarium sealant to prevent escapes and minimize draughts.

Buy or make removable shallow glass or plastic trays to fit the base of the vivarium. Ensure that the top edges of the glass are smooth.

The sliding glass doors should open and close easily. Rub candle wax along the top and bottom edges to help reduce the friction.

Affix fine wire mesh to the outside of the ventilation holes using tacks or strong glue. Make sure there are no small escape points.

**Guide to vivarium dimensions for one to four adult lizards**

| Size of lizard | Length | Height | Depth |
| --- | --- | --- | --- |
| Up to 13cm(5in) | 45cm(18in) | 30cm(12in) | 30cm(12in) |
| 13-38cm(5-15in) | 75cm(30in) | 45cm(18in) | 38cm(15in) |
| 38-60cm(15-24in) | 90cm(36in) | 60cm(24in) | 60cm(24in) |
| over 60cm(24in) | 120cm(48in) | 75cm(30in) | 60cm(24in) |

For arboreal lizards, substitute height for length and vice versa.
See pages 32-39 for advice on furnishing the vivarium.

cover the wood with adhesive plastic film or to paint it with a sealant, such as the waterproofing paint used in concrete pool construction. Chipboard laminated with plastic or coated in melamine is another, albeit more expensive, alternative.

If you intend using very high wattage lamps or ceramic heaters, you may wish to coat the vivarium with a flame-retardent paint available from DIY stores. When completely dry, the paint is harmless to lizards. Secure a strong metal gauze or custom-made plastic air vent to the outside of the vivarium. The size of the ventilation hole should be in proportion to the size of the vivarium. If it is too small, stale air will accumulate; too large and it will create draughts that could give rise to cold-related infections in the lizards.

There are various ways of incorporating a door into the vivarium. One method is to attach a wooden-framed glass door to one side with hinges. Such a door is easy to clean but, when open, could prove an irresistible escape route, particularly for small, agile lizards. Sliding glass doors on plastic or

*Below: Always use sturdy branches and logs as decoration, and do not balance pieces precariously. This male skink, Eumeces inexpectatus, in breeding colours, is resting safely on a strong branch.*

wooden runners reduce the risk of escape, but also restrict access, making both the vivarium and the inside of the door more difficult to clean. If you opt for this system, smooth the edges of the glass and wax them to reduce resistance on the runners. Glass is better than perspex or acrylic because it is less likely to scratch and easier to clean.

Once the vivarium is assembled, check that there are no gaps around any holes drilled for ventilation grills or electric cables. Fill in all the joints and corners with silicone sealant, available from most pet and aquarist shops. This not only helps to stop draughts and potential escapes, but also prevents the build-up of debris, bacteria and parasites in the cracks. Furthermore, it makes the vivarium leakproof when it is washed down.

The vivarium will be easier to clean if you install removable glass trays at the bottom. These are simple to make or you can buy them ready made from an aquarist shop.

### Keeping the vivarium clean

Good hygiene is a major factor in preventing disease in captive lizards. Clean out the vivarium every 7-10 days, depending on the species. This entails washing or replacing the substrate and washing or disinfecting any decorative branches, rocks, water bowls or hide-boxes. An effective but cheap disinfectant is a 3 or 4 percent sodium hypochlorite solution (household bleach) but be sure to rinse the items thoroughly in running water after use.

If plants are incorporated into the set-up, examine the roots closely for signs of potentially dangerous pests. Keeping plants in plastic pots makes this task much easier.

On a day-to-day basis, inspect all vivariums for faeces, shed skin, uneaten food or even dead lizards, all of which should be removed immediately and not left until the next regular maintenance is due.

### Siting the vivarium

Place the vivarium in a light position, but never in full sunlight, otherwise both the container and its occupant will overheat. Temperate species require no extra heat if they are in a partially sunny position. During the cooler months, you may choose to hibernate them, but it is safer to overwinter juveniles in warm surroundings.

Thermostatically controlled plant propagators that enable you to monitor both temperature and humidity are ideal for warm temperate and tropical species. Another strategy is to place several small containers in the range of a light bulb or fluorescent tube, thus reducing electricity costs.

### Vivarium wear and tear

Checking for wear and tear is particularly important for DIY vivariums, especially wooden designs, where humidity or spilt water seeping into joints can cause warping. The resulting gaps provide an escape route for smaller lizards as well as for live foods, such as crickets or mealworms. Recondition all homemade vivariums at least once a year, sealing joints, treating or recoating wood with waterproofing paint and tightening screws, light fittings or door hinges.

Below: *A vivarium can become a focal point if it is well furnished with plants. The lizards will appreciate the cover, particularly if there are glass sides.*

# HEATING AND LIGHTING

Accurate control of heating and lighting is possibly the most important yet least understood aspect of maintaining lizards in captivity. If applied properly, heat and light will stimulate the appetite, promote overall health and invigorate the reproductive activity of the lizard. Many different heating and lighting systems are in popular use today, some more efficient than others. Before choosing one particular arrangement, consider the precise needs of the species you intend to keep. This is especially important if you contemplate housing several different types of lizard together, as their requirements may differ.

Above: *Ceramic heaters are available in many shapes, sizes and wattages, but they can be quite expensive.* *Make sure you also buy the appropriate holders. Remember that these heaters can become very hot.*

## Heating the vivarium

A vivarium can be heated either by a single incandescent light or by an elaborate combination of thermostats, heater pads and ceramic spotlamps, depending on your budget, on what is best for the species and on the type of vivarium. It does not make sense, for example, to install an expensive heating arrangement for a slow worm or a burrowing species; on the other hand, a 40-watt light bulb will not be adequate for, say, a species originating from equatorial Africa.

## Heating lamps

Although a single incandescent light bulb may seem a very simple option, it will in fact suit many of the 'easier' species, particularly

Below: *It is a good idea to attach a reliable thermometer to the* *vivarium to give an indication of air temperatures.*

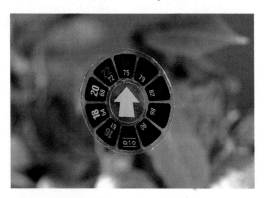

temperate lizards. It is economical to set up and run and is adequate for temporary accommodation, such as quarantine quarters. Match the power of the lamp to the size of the vivarium (see page 36-37), and opt for a spotlamp rather than an ordinary bulb as it will provide a 'hot-spot'. The positioning of the lamp is important. Place the lamp towards one side of the vivarium and tilt it so that it shines onto a flat rock or log underneath. This will enable the lizard to bask under the hot-spot until its body temperature has reached an optimum level and then hunt prey or retire to a cooler part of the vivarium.

A ceramic lamp maintains heat at a higher temperature and is suitable for the larger vivarium or for species requiring hotter conditions. Try to obtain a model equipped with special protective reflectors, which will prevent the vivarium from overheating, and position it in the same way as a spotlamp.

The obvious disadvantages of heater lamps are that the exact temperature of the vivarium will be determined partly by outside ambient room temperatures, and that extra heating is not provided at night, unless the room is centrally heated (and the heating

## Heating and lighting the basic vivarium

*Attach fluorescent lighting tubes and other lights to the roof of the vivarium before joining it to the side and back panels.*

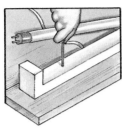

*Holders for ceramic heaters or spotlights should be firmly secured. If they work loose, they could become a fire hazard.*

*Slack cables are both dangerous and unsightly. Fasten them to the inside of the roof and to the sides of the vivarium.*

*To prevent the risk of burns, encase all spotlights and ceramic heaters with a protective lightweight wire mesh cage.*

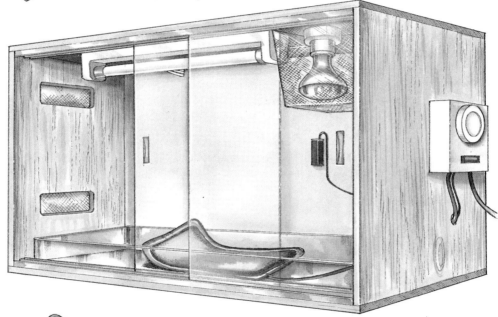

*Direct the cable from the heater pad into the thermostat control box, which should be accessible from outside the vivarium.*

*Set the thermostat controls in accordance with the instructions, bearing in mind the requirements of the species to be housed.*

*When light fittings are installed, nail the roof into position, taking care that wires fit through groove at top of back panel.*

*Slot glass doors into runners. Check the wiring and test all electrical appliances to make sure they are functioning safely.*

Above: *In order to regulate its body temperature in hot conditions, this* Agama agama *has lifted its body from the heat-retaining rocks and opened its mouth.*

left on overnight) or the lamp is left on permanently. To leave a lamp on continuously would be unnatural and would disrupt the daily and long-term habits of the lizard, so a heatlamp-only vivarium is limited to those species able to withstand cooler night temperatures.

Higher wattage spot and ceramic lamps can become extremely hot, so enclose them in a wire mesh box and do not allow any branches or rocks to touch them. Direct the heat from the lamp onto rocks, not onto the vivarium wood and ensure that water cannot come into contact with a lamp by positioning the waterbowl at the side furthest away from the lamp. If you spray the inside of the vivarium to create a fine mist, always direct the spray away from lamps.

### Heater pads

Heater pads for vivariums are widely available in a variety of different shapes and wattages. They range from simple plastic or metal platelike devices to plastic heaters simulated to resemble rocks. Alternatively, heater pads used in beer-making will work equally well. Whichever design you choose, check it thoroughly to ensure that it is safe and waterproof. Pads are very economical to run, especially in conjunction with a thermostat, and are essential as a back-up if the room temperature should accidentally fall below the critical level required by the species in the vivarium.

Platelike heaters and horticultural soil-warming cables can be concealed beneath the substrate, the depth depending on the power and size of the pad. For example, a 20-watt,

20cm(8in) square pad would function efficiently at a depth of between 5 and 10cm(2-4in). It is possible to achieve a high level of humidity in a tropical, planted vivarium using this method of heating, but take care not to dry out plant compost. Rock-type heaters work well for thigmothermic species, but choose a low wattage model, as some makes are liable to overheat.

Do not use a heater pad as the main source of heat, because it does not provide a natural basking area and affords few cool spots if the lizard overheats. Bear in mind that although a heater pad controlled by a thermostat may achieve an ideal air temperature, the substrate may become too hot.

### Thermostats

A thermostat regulates the warmth of the vivarium by automatically switching off the heat source when the vivarium reaches a preset temperature. It is not a good idea to use a thermostat to control the heat from an incandescent lamp, as constantly switching the light on and off is liable to stress the lizard, with potentially fatal consequences. At the very least, the lizard will have no perception of day and night, which will result in diminished reproductive behaviour. Thermostats are much better employed in controlling heater pads and are particularly recommended for higher wattage pads, which may become too hot if not regulated. In this case, position the thermostat sensor so that it reads the temperature of the substrate, not of the air.

### Lighting the vivarium

In nature, the sun is the source of both light and heat. In the vivarium, a heatlamp performs a similar function. Although such lamps are adequate for, say, the subterranean species, the terrestrial, heliothermic species need something more to maintain their health and encourage breeding. This 'extra ingredient' comes in the form of ultraviolet light, emitted by suitable bulbs or balanced daylight fluorescent tubes. Ultraviolet light has important germicidal properties and plays an essential part in the mobilization and synthesis of vitamin $D_3$ and its precursors, calcium and phosphorous. Vitamin $D_3$ is essential for maintaining a healthy appetite, skin and skeletal

## Heating and lighting options

Timers or time switches are obtainable from most electrical appliance stores. Simply plug them into the mains and connect the vivarium lighting plug into the same socket. The switch-on and off times are then set via a clock. The actual time settings selected depend on the lizard's country of origin, the time of year and whether you are hoping to breed the lizards.

Heater pads incorporate a metal heating coil protected by a plastic or wooden cover. Most of them are fitted with a thermostat that switches off the pad at a preset temperature. More expensive models have a variable adjuster, but if this is not supplied, install an external thermostat in a suitable position to monitor the temperature in the vivarium.

Light bulbs are available in a variety of wattages and colours. Normal incandescent bulbs and spotlamps are ideal for diurnal and basking lizards, whereas the orange light and blue spot lamps will suit nocturnal species not accustomed to harsh light. Direct a blue spotlamp onto a rock for thigmothermic lizards that seek out heat-retaining rocks at night.

Fluorescent tubes are ideal for lighting a vivarium, particularly one designed to house temperate lizards that do not naturally bask. The tubes are long-lasting and economical in use, and available in a range of sizes. For healthy plant growth, choose a balanced daylight fluorescent tube that produces the spectrum of wavelengths found in sunlight.

Above: *Fluorescent tubes, both lit. The powertwist (top) emits more UV light and is ideal for both wide and tall vivariums.*

## Black UV lights

High UV-output, black lights in the form of a tube or bulb are suitable for large vivariums and popular in the US. (Use BL - 'black light' - bulbs, not BLB - 'black light bright' bulbs, which could be harmful.) BL bulbs emit very little light and can be used in conjunction with a heat lamp or fluorescent tube. The life expectancy of all UV-emitting tubes is about 20,000 hours, or about four to five years of daily use, which probably outweighs the disadvantage of their initial high cost. UV-emitting bulbs, although less expensive, have a considerably shorter life expectancy than tubes (around 4,000 hours) and emit weaker UV light. Their only advantage is that they give out more heat than the tube.

pages 46 and 60). Ultraviolet light can also have a remarkable effect if you want to encourage your lizards to breed (see also the table on page 31).

The fluorescent tubes range in length from 45 to 240cm(18-96in) and are rated at 15 to 125 watts (the longer the tube the greater the wattage). Perhaps the most effective tubes are those with a powertwist (literally a twist in the glass tube that increases the surface area of light-emitting phosphors). They emit over 15 percent more light than a normal tube of the same wattage and also have a slightly longer life.

## Light intensity

It is important to achieve the correct intensity of light, especially if you wish to encourage plant growth. Remember that in a tall planted vivarium the light intensity will diminish rapidly from the light source at the top of the vivarium down to the base.

## The photoperiod

Photoperiod, or the length of daylight, indirectly plays a very important part in the lizard's behavioural pattern. Almost all species of lizard experience a change in photoperiod throughout the year; those from temperate regions encounter the greatest range, equatorial species the smallest. In the vivarium, it is possible to simulate not only the natural photoperiod (and the periodical

## Guide to preferred temperatures

| Type of vivarium | Daytime temperature range Min/Max | Night-time temperature range Min/Max |
|---|---|---|
| Cool temperate | 18-27°C(64-80°F) | 10-18°C(50-64°F) |
| Warm temperate | 21-32°C(70-90°F) | 15-21°C(60-70°F) |
| Sub-tropical | 27-32°C(80-90°F) | 21-24°C(70-75°F) |
| Tropical | 27-32°C(80-90°F) | 24-29°C(75-85°F) |
| Semi-desert/desert | 29-38°C(85-100°F) | 10-21°C(50-70°F) |

## Guide to recommended photoperiods

| Type of vivarium | Average photoperiod (hours per day) | | | |
|---|---|---|---|---|
| | Spring | Summer | Autumn | Winter |
| Cool temperate | 12 | 14 | 12 | 8 |
| Warm temperate | 10 | 14 | 12 | 10 |
| Sub-tropical | 12 | 14 | 12 | 10 |
| Tropical | 14 | 16 | 14 | 12 |
| Semi-desert/desert | 12 | 14 | 12 | 10 |

changes that the lizard would encounter in the wild), but also to create unnatural photoperiods to encourage breeding. The photoperiod required by a lizard depends both on its place of origin and the time of year. The table provides a general guide to seasonal photoperiods, but these obviously vary according to the individual species.

### Time switches
Time switches are essential in order to achieve the correct photoperiod for each lizard species. In the last few years, some excellent time switches have appeared on the market; some dim the light gradually so as not to plunge the lizards into sudden darkness (and vice versa). Others are controlled digitally, providing the facility to preset different time ranges over a period of 1-14 days. A combination of programming and override means that the vivarium can be heated and lit even when the owner is away for up to two weeks.

### Inspecting electrical appliances
Accidents do happen from time to time, and faulty wiring or heating elements can have a potentially devastating effect on the vivarium. To keep these risks down to a minimum, always buy good quality, guaranteed equipment. Possibly the most useful aid for the herpetologist is a circuit breaker that regulates the flow of electricity, immediately switching off at the first sign of danger, such as a blown fuse or electrical

fault. Once incorporated into the set-up, keep a close eye on the temperatures given off by lights and heaters and check heating and lighting periods regulated by time-switches. If the equipment does not function correctly, then consult a qualified electrician or return it to the shop.

Below: *An attractive, tall planted vivarium, lit from above. If the room temperature remains warm, there will be little heat variation in the vivarium. Slightly warmer air may circulate towards the top, depending on the power of the light.*

# FURNISHING THE VIVARIUM

Having constructed the basic vivarium and considered the heating and lighting options, the next step is to 'customize' it to suit the needs of the species you wish to keep. In this section we consider the role of plants in the vivarium and then look in more detail at different set-ups that can be adapted quite easily as required.

### Plants in the vivarium
Live plants play an important role in creating an attractive and realistic setting for lizards, as well as providing them with shelter. Remember that some lizards are herbivorous, so make sure the plants are not poisonous or use plastic vegetation. Set live plants directly into a suitable growing medium or, preferably, sink the plant pots into the substrate or hide them behind rocks and logs. Check the roots for any potentially harmful pests. The plants will require the appropriate lighting, as described on the following pages.

### The subterranean vivarium
Amphisbaenids, or worm lizards, live almost exclusively underground, only coming to the surface after heavy rainfall, or to feed or at nightfall. Although they are shy and secretive, their nocturnal habits make them highly interesting and they will occasionally breed in captivity. An all-glass vivarium is ideal for these species. A pair of the largest species, such as adult slow worms, *Anguis fragilis*, will require a vivarium measuring 90x60x60cm(36x24x24in).

The first step is to select a suitable substrate. A 10-20cm(4-8in)-deep layer of a loose, dry material, such as sterilized chipped bark is one possibility. Broken crocks or cork bark on the surface provide added decoration, along with a small water bowl. If the vivarium is to house temperate species, such as slow worms, the heating can take the form of a low-power spotlight (25-40watt is sufficient). For more exotic species, you may need to add a thermostatically controlled heater pad under the substrate. UV light is not needed in this vivarium. Add a tight-fitting lid with adequate ventilation and change the substrate at least once a month.

Below: *Subterranean lizards, such as* Anguis fragilis, *are at home in slightly humid conditions in a loose substrate.*

## Completing the vivarium decor

If a heater pad is incorporated into the vivarium, cover it with newspaper to act as insulation and to absorb moisture.

Use a hammer drill to bore large hiding holes in soft rock, such as tufa. If the holes are too small, they may harbour pests.

Water dishes should be deep enough to allow the lizards to bathe, but not so deep that they drown. Change the water every day.

Spotlamps and ceramic heaters can become very hot. Cover them with aluminium mesh, particularly if there are branches nearby.

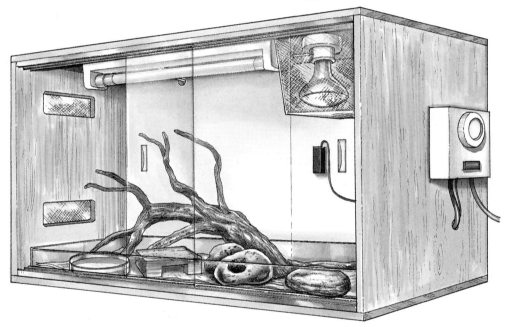

Disinfect all branches and rocks in a bleach solution and rinse well before placing in the vivarium. Remember to wear rubber gloves.

A simple box-type hide is easy to make, using a hard wood coated in waterproof paint. The lizards will appreciate a choice of retreats.

Pieces of tufa rock can be firmly glued together to create a sturdy structure that will not topple over and crush the lizards.

Place a large flat rock under the hot spot. The warmth retained in the rock creates a welcome heat source for the thigmothermic species.

## The terrestrial vivarium

Under this general heading there are three distinct categories, each designed to suit the requirements of different lizard species.

**The savanna or desert** arrangement is suitable for lizards that experience high daytime temperatures and little in the way of rainfall in their natural habitat. It can look very attractive and requires less attention than some other set-ups.

First prepare the substrate, which should be 2.5-5cm(1-2in) deep, using one of the many grades of gravel available from aquarist shops. It is both easy to clean and replace. Wash the gravel in a 3 percent solution of sodium hypochlorite (household bleach), then thoroughly rinse it in running water and dry it. Avoid using sand in a vivarium, unless you intend housing sand skinks that enjoy 'swimming' just underneath the surface. It is likely to clog the glass door-runners, stick to food particles and fall into drinking water, where it is easily ingested by the lizard, causing respiratory and digestive complications.

Decorate the remainder of the vivarium with granite and limestone rocks, cork bark, dead branches and driftwood. The lizards will take refuge under these features, so make sure they are completely stable and do not balance pieces precariously on top of one another. A collapse could be fatal.

Tufa rock - a calcareous rock, chemically very similar to coral sand - is another popular decorative material, widely available and easy to work with. Deep holes bored into it with a drill or chisel make ideal hideaways for small lizards. A ceramic water bowl completes the decoration.

The heating in such a vivarium will depend on the species housed in it. A high-wattage spotlight or ceramic heater will satisfy most requirements, but make sure there are also enough cool spots and sufficient cover so that the lizards can escape the intense heat. Most diurnal species - and any plants in the vivarium - will require UV light as well. Tender lizard species will benefit from a heater pad switched on at night, although many can endure a significant drop in temperature as long as there are sufficient hiding places.

Suitable plants for this vivarium include the less spiny cacti and succulents, such as *Astrophytum*, *Myrtillocactus*, *Stapelia*, *Haworthia*, *Aloe* and *Yucca* species. Avoid the poisonous *Euphorbia* and *Lophophora* species. Water the plants once a fortnight and check the roots and soil for potentially hazardous organisms. The whole vivarium can be

Below: *The horned toad,* Phrynosoma cornutum, *prefers a dry sand or gravel substrate, where the temperature can reach the high levels found in its natural home.*

Left: *Many species of lizard, such as anoles, enjoy foraging in trees and shrubs for food and refuge, or to bask.*

Overleaf: *This table is a general guide to heating and lighting requirements in a range of vivariums. The final choice of equipment will depend on the lizard species to be housed, the ambient room temperatures and whether the vivarium is located in a bright or dimly lit site.*

lightly misted in the evening using a plant sprayer to create an effect of dew forming.

**The woodland or forest** vivarium will suit many terrestrial and semi-arboreal species from temperate, warm temperate and tropical regions. It differs from the desert vivarium in that ambient temperatures are more constant and the humidity is higher.

Coarse gravel, chipped bark or sphagnum moss peat, laid to a depth of 5-10cm(2-4in), are all suitable substrate materials, but they require frequent sterilizing or regular replacement so that mould and algae do not develop. To sterilize bark or peat, place it in a closed biscuit tin and heat it in the oven at 100⁰C(212⁰F) for one hour or so.

Embed rocks, bark and logs in the substrate, ensuring that they are all well anchored. The size of the water bowl depends on the species in the vivarium. Iguanas, basilisks, teiids and agamas of the genus *Hydrosaurus* all enjoy semi-aquatic activities, but other lizards may drown in too much water.

The heat from a warm room, say 20-24⁰C(68-75⁰F), will suit most temperate lizards, but the more exotic species will require additional heating at night, best provided by a heater pad buried in the substrate. A pad is also useful if you want to create a more humid atmosphere. You will need a spotlight (select the appropriate wattage according to the species in the

vivarium), but direct it away from live plants. A UV tube is essential in all cases. Control both heat and light with a time-switch.

Larger lizard species will devastate a planted vivarium, so confine the decoration to imitation plants, some of which look very realistic. Smaller lizards are not destructive and you can introduce some attractive plant life into their vivarium. The plants will vary according to the climate enjoyed by the lizards. In a temperate vivarium you might consider the castor oil plant (*Fatsia*), ferns (*Phyllitis* and *Cyrtomium*), geraniums (*Pelargonium*), mother-in-law's tongue (*Sansevieria*), and ivy (*Hedera*). All grow well at lower temperatures if regularly watered.

The majority of houseplants are suitable for a warm temperate or tropical environment, but goosefoot (*Syngonium*), the parasol plant, (*Heptapleurum*), *Caladium* and *Aglaonema* all seem better able to withstand higher temperatures.

A tropical vivarium will need daily misting, but take care that excessive moisture does not lead to fungal growth on the sides of the vivarium.

**The tall planted vivarium** is basically the same as the previous vivarium, but specifically designed for arboreal species, such as day geckos, *Phelsuma* species, chameleons and certain iguanids.

Provide the same lighting as for the woodland vivarium, but add a UV tube with

## Guide to heating and lighting a range of vivariums

| Type of vivarium | Conditions/temp. range | Size of vivarium (LxDxW) |
| --- | --- | --- |
| Subterranean | Temperate 10-27°C(50-80°F) | up to 90x38x30cm(36x15x12in) over 90x38x30cm(36x15x12in) |
| | Tropical 24-32°C(75-90°F) | up to 90x38x30cm(36x15x12in) over 90x38x30cm(36x15x12in) |
| Savanna / Desert | Desert 10-38°C(50-100°F) | 60x38x30cm(24x15x12in) 90/120x45x30cm(30/48x18x12in) 150x45x38cm(60x18x15in) |
| Woodland / Forest | Temperate 10-27°C(50-80°F) | 60x38x30cm(24x15x12in) 90/120x45x30cm(36/48x18x12) 150x45x38cm(60x18x15in) |
| | Tropical 24-32°C(75-90°F) | 60x38x30cm(24x15x12in) 90/120x45x30cm(36/48x18x12in) 150x45x38cm(60x18x15in) |
| Tall Planted | Temperate 10-27°C(50-80°F) | 38x60x30cm(15x24x12in) 45x90x30cm(18x36x12in) 45x120x38cm(18x48x15in) |
| | Tropical 24-32°C(75-90°F) | 38x60x30cm(15x24x12in) 45x90x30cm(18x36x12in) 45x120x38cm(18x48x15in) |
| Simple / Hygienic | Temperate 10-27°C(50-80°F) | 60x38x30cm(24x15x12in) 90/120x45x30cm(36/48x18x12) 150x45x38cm(60x18x15in) |
| | Tropical 24-32°C(75-90°F) | 60x38x30cm(24x15x12in) 90/120x45x30cm(36/48x18x12in) 150x45x38cm(60x18x15in) |

Heating and lighting

40-watt incandescent bulb
60-watt incandescent bulb

40-watt bulb and 15-watt underground cable
60-watt bulb and 15-watt underground cable

45cm(18in) 15-watt balanced daylight tube and 40-watt spotlamp. 15-watt heater pad (at night)
45cm(18in) 15-watt balanced daylight tube and 60-watt spotlamp. 15-watt heater pad (at night)
75cm(30in) 25-watt balanced daylight tube and 100-watt spotlamp or ceramic heater. 15-watt heater pad (at night)

45cm(18in) 15-watt balanced daylight tube and 40-watt spotlamp or bulb
45cm(18in) 15-watt balanced daylight tube and 60-watt spotlampor bulb
75cm(30in) 25-watt balanced daylight tube and 80-watt spotlamp or bulb

45cm(18in) 15-watt balanced daylight tube and 40-watt spotlamp or bulb. 15-watt heater pad
45cm(18in) 15-watt balanced daylight tube and 60-watt spotlamp or bulb. 20-watt heater pad
75cm(30in) 25-watt balanced daylight tube and 80-watt spotlamp or 100-watt bulb. 25-watt heater pad

40-watt spotlamp or bulb or 60-watt UV bulb and 15-watt heater pad as backup
40-watt spotlamp/incandescent bulb and 45cm(18in) 15-watt balanced daylight tube. 15-watt pad as backup
60-watt spotlamp/incandescent bulb 45cm(18in) 15-watt balanced daylight tube. 20-watt heater pad as backup

40-watt spotlamp/incandescent bulb or 60-watt UV bulb. 20-watt heater pad
60-watt spotlamp/incandescent bulb and 45cm(18in) 15-watt balanced daylight tube. 20-watt heater pad
60-watt spotlamp/incandescent bulb and 45cm(18in) 15-watt balanced daylight tube. 25-watt heater pad

40-watt spotlamp/incandescent bulb and 45cm(18in) 15-watt balanced daylight tube
60-watt spotlamp/incandescent bulb and 45cm(18in) balanced daylight tube
100-watt spotlamp/incandescent bulb or ceramic heater and 75cm(30in) 25-watt balanced daylight tube

40-watt spotlamp or 60-watt incandescent bulb and 45cm(18in) 15-watt balanced daylight tube. 15-watt heater pad (at night)
60-watt spotlamp/incandescent bulb and 45cm(18in) 15-watt balanced daylight tube. 20-watt heater pad
100-watt spotlamp/incandescent bulb and 75cm(30in) 25-watt balanced daylight tube. 25-watt heater pad

a power twist or a black UV light. The light from an ordinary tube may not reach the lower part of the vivarium and the plants growing there.

Nearly all arboreal lizards originate from warmer climes, so this is a good opportunity to grow some interesting plants. You might consider one of the epiphytic bromeliads or a staghorn fern (*Platycerium*), which look very effective when attached to upright logs and branches using silicone sealant. Plant Spanish moss, (*Tillandsia usneoides*), around the base. Tall-growing plants, such as the dragon tree, (*Dracaena*), yuccas or ornamental figs, (*Ficus*), are also very appealing.

Flat wooden or plastic platforms, secured to the sides of the vivarium with silicone sealant, are ideal for supporting a small water bowl and items of food. Smaller lizards will seek out cut sections of hollow bamboo cane or plastic piping in the corners of the vivarium as hiding places or as somewhere to deposit their eggs.

**The simple vivarium**

All the vivariums described so far require constant inspection for signs of parasitic infestation or disease, and their substrates must be cleaned out or regularly replaced. The two remaining vivariums described here require little in the way of day-to-day maintenance and are ideal for hobbyists who spend little time at home.

Although not as attractive as a planted vivarium, the simple set-up is probably the most popular one, especially if you wish to accommodate several different species in separate vivariums. Cover the base with dry disposable material, such as newspaper or paper towelling, or with washable sponge or cotton lining. Heating and lighting requirements depend entirely on the species being kept. This type of vivarium will accommodate most systems, ranging from incandescent bulbs to ceramic spots. If the substrate is a combination of newspaper and a heater pad, select a low-wattage pad and lay down several thick layers of newspaper.

Keep the decor sparse, consisting of a water bowl, a few branches and a wooden box shelter large enough for all the vivarium occupants. Many lizards adapt well to this artificial arrangement, especially juvenile or captive-bred species. Another advantage of

Above: *A tall planted vivarium needs UV lighting to encourage healthy plant growth. It is ideally suited to anoles and day geckos.*

Below: *The outdoor vivarium is intended to simulate the lizards' natural habitat. A sunny location is, therefore, essential.*

## Building an outdoor vivarium

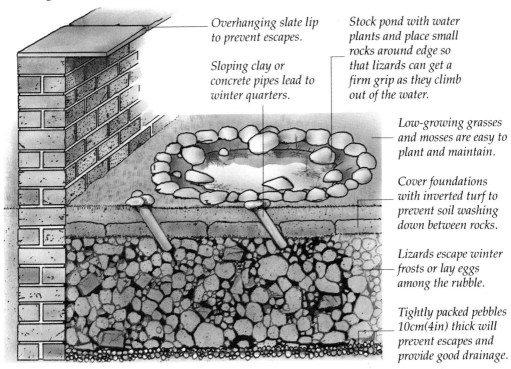

*Overhanging slate lip to prevent escapes.*

*Sloping clay or concrete pipes lead to winter quarters.*

*Stock pond with water plants and place small rocks around edge so that lizards can get a firm grip as they climb out of the water.*

*Low-growing grasses and mosses are easy to plant and maintain.*

*Cover foundations with inverted turf to prevent soil washing down between rocks.*

*Lizards escape winter frosts or lay eggs among the rubble.*

*Tightly packed pebbles 10cm(4in) thick will prevent escapes and provide good drainage.*

this design is that it enables you to keep a lookout for signs of disease or pregnancy.

### The outdoor vivarium

The final option considered here is an outdoor vivarium. This could be an open, walled enclosure, often called a reptiliary, or an 'oversized' glass vivarium, such as a greenhouse. Both provide an opportunity to study the natural behaviour of temperate lizard species and are very easy to maintain.

A sunny, well-drained position is essential for a reptiliary. Start by laying deep foundations - minimum 60cm(24in) but 90-107cm(36-42in) is better, with plenty of 'rubble' at the base. This will provide good drainage and adequate hiding or hibernation sites. If it is to be used exclusively for lizards, install a small pool and plant a miniature version of the lizards' natural habitat around it, remembering that the vegetation should not be allowed to grow too tall. This set-up is only suitable for smaller, hardy lizards, such as *Lacerta*, *Anguis* and *Podarcis* species.

To prevent lizards escaping, fix a strip of clear acrylic sheeting or some roofing tiles to the top bricks, creating an overhang of 20-

25cm(8-10in). Keep out rats, hedgehogs and cats by covering the whole vivarium with wire mesh.

Whatever precautions you take, some escapes from an open enclosure are inevitable. A greenhouse is much more secure and you can introduce more exotic species, especially if you install a suitable heater for the wintry seasons. Be sure to obstruct any possible points of escape, such as vents, doors and joints. Realistically, the greenhouse is suitable only for small to medium-sized ground dwellers, but even then the choice of species is almost limitless.

Hardy, small-growing shrubs and plants will thrive in the greenhouse, especially ferns, heathers (*Erica*), dwarf conifers and various grasses.

The only source of heating and lighting in an outdoor vivarium is the sun, but this can be a problem in the greenhouse if it becomes too hot. The greenhouse will need shading in summer and, if you intend to keep more exotic species, an electric fan heater at night or on overcast days. A ceramic spotlight plus fan heater will provide adequate heating during the colder months.

# FEEDING LIZARDS

To maintain a lizard in good health, you must offer it a balanced and varied diet that will provide the animal with all the essential elements it would find in the wild. When first introduced to a captive environment, wild-caught lizards may refuse food for a short period until they settle down. If this fasting continues for more than three to four weeks (or ten days in very small lizards), then the lizard may be sloughing, gravid or sick in some way. See your veterinarian for a diagnosis; if the lizard is deemed perfectly healthy, you may need to force feed it (see page 47). Newly obtained, captive-bred lizards may also refuse food and, again, this may be caused by the stress of moving, gravidity or sickness, although the latter is rare. Some species (see page 41) are notoriously difficult to feed at any stage and are best avoided.

## Catching and eating food

Smaller lizards, such as lacertids, rely on their keen eyesight and sense of smell to stalk their prey - usually insects - and then speedily grasp the food in their jaws. Some lacertids enjoy a certain amount of fruit as well as carrion in their diet and locate this with the Jacobson's organ (see page 15).

Chameleons are indisputably the expert catchers, having excellent vision and a specialized extendible, muscular, sticky tongue. Very rarely do they miss their target. Agamids also have muscular, sticky tongues, with which they 'lap up' ants or termites. Larger lizards, such as tegus, monitors and larger skinks, use the Jacobson's organ to locate small mammals, eggs, carrion or fruit. Semi-burrowers and subterranean species with poor eyesight also rely on the Jacobson's organ to track down earthworms, slugs and snails. Geckos and small iguanids react to the movement of their prey, whereas large iguanids are sensitive to the smell and colour of their plant food.

Below: *The giant zonure,* Cordylus giganteus, *is entirely carnivorous, and is even known to eat carrion in the wild.*

No lizard is able to 'taste' its food to the same degree as a mammal, so it simply aims to subdue and swallow the food as quickly as possible. Many lizards violently shake their prey and tear or chew it into pieces of a suitable size. Be sure to protect your hands when feeding lizards, especially the larger lizards, which can deliver a nasty bite with their powerful jaws.

Lizards can be divided into five categories according to their feeding habits: specialist feeders, insectivores, carnivores, herbivores and omnivores, and between them they take an enormous variety of foods.

### Coping with specialist species
Specialist feeders accept only one type of food - a particular insect, for example - and are thus extremely difficult to maintain in captivity. You may be able to train a specialist feeder to accept an alternative food, especially if it is very hungry, but this approach is not always successful and you would do well to avoid this type of lizard. Examples of specialist feeders include,

Above: *Crickets are the commonest form of food for the majority of smaller lizards, such as* the insectivorous sand gecko, Teratoscincus scincus, *shown here. They are easy to breed.*

*Moloch horridus*, which feeds solely on ants and termites, some horned lizards, *Phrynosoma*, which also specialize on ants and the Caiman lizard, *Dracaena guianensis*, which feeds exclusively on freshwater snails.

### Feeding insectivorous lizards
The majority of smaller lizards are insectivorous and some of the large species enjoy insects in their diet. Fortunately, insects and other invertebrates are widely available. You can catch them yourself or buy them in small or large quantities at a very reasonable price from specialist shops or mail-order companies. Specialist laboratories are a particularly important source of supply in the winter, when invertebrates can be in short supply. Here, we look at the most commonly available invertebrates that make suitable food for insectivorous lizards.

**Crickets** (*Acheta domestica* or *Gryllus bimaculatus*) are probably the most widely used food for lizards, as they are relatively cheap, easy to breed and their variation in size suits small and larger lizards alike. Hatchlings are ideal for juvenile lizards, as they are only 1-2mm(0.04-0.08in) long and have a soft body. Adult crickets measure up to 20mm(0.8in) and are ideal for larger lizards, but avoid offering large female crickets, distinguishable by the long spiky ovipositor used for egg laying. This protrudes from the female's abdomen and may penetrate the lizard's oesophagus or stomach lining, causing infection.

House the crickets in an aquarium covered with a fine mesh, such as muslin, and keep them at a temperature of 21-27⁰C(70-80⁰F). Line the base with newspaper covered with a layer of cardboard tubes or crumpled newspaper to increase running space. (These will need replacing at regular intervals.) Add two small containers, such as coffee jar lids, one for food - chopped apple, carrots or lettuce - and the other for water - supplied on cotton-wool soaked in 40mls of water enriched with 3 or 4 drops of multivitamin liquid. Provide another small container filled with damp sand for the females to deposit their many eggs. Remove these eggs at intervals and incubate them at about 24⁰C(75⁰F) until they hatch, three or four weeks later.

**Locusts** (*Locusta migratoria*) are suitable for larger lizards, such as tegus and monitors and, although they tend to be quite expensive, they do provide plenty of nourishment. Juvenile locusts, or hoppers, are soft bodied, make excellent food for many small lizards and are widely available from laboratories. You can keep locusts in the same

### Breeding crickets in an aquarium

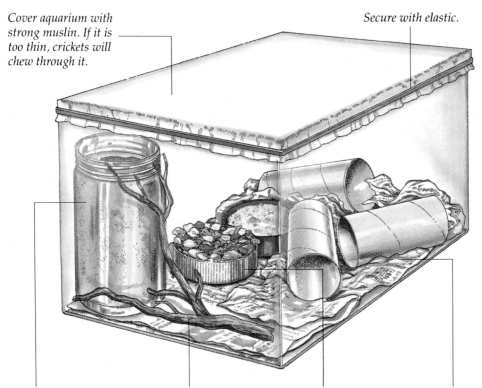

*Cover aquarium with strong muslin. If it is too thin, crickets will chew through it.*

*Secure with elastic.*

*Female crickets lay eggs in jar filled with damp sand. Do not allow sand to dry out.*

*Place a branch in the aquarium so that the crickets can reach the sand in the glass jar.*

*Without suitable food, the crickets will eat each other. Provide fresh water regularly.*

*Crumpled paper and cardboard tubes add running space so more crickets can be housed.*

way as crickets, but they will need warmer conditions (around 32°C/90°F) and more spacious quarters in which to breed. (Although they will survive at lower temperatures, they will grow very slowly.) However, locusts can be expensive to breed and are better bought just as a food supply.

**Mealworms** (larvae of the flour beetle, *Tenebrio molitor*) are an easy-to-breed food supply for lizards and particularly relished by agamids and lacertids. Do not make them the only food for lizards, however, as they contain little in the way of calcium, but they are ideal for 'fattening up'. Rear them at room temperature in ice-cream tubs half filled with loose, flaky bran, on which they feed. Mix powdered cuttlefish bone with the bran to increase the amount of calcium, and place a slice of apple or potato on the surface for moisture and food, replacing it as it dries.

Breeding is easy; the adult beetles lay hundreds of eggs, which hatch after six to eight weeks and reach pupae stage 16-20 weeks later. By starting cultures progressively, you will have a constant supply of larvae throughout the year. The mealworms moult at regular intervals, after which they are soft and white. On hardening, the exoskeleton is quite indigestible, so offer the freshly moulted mealworms to juveniles or lizards with weaker jaws.

Above: *Buy locusts in small quantities from specialist suppliers. They are not cheap to buy, but are an ideal food for larger lizards.*

Below: *Mealworms are available in three different ascending sizes: normal, giant (zophobas) and jumbo (buffalo worms).*

Left: *Agile lizards, such as the common wall lizard,* Podarcis muralis, *can easily catch houseflies. For more docile species, cool the fly down in a fridge for five minutes to slow it down or remove the wings.*

Right: *Some captive-bred lizards, such as* Ophisaurus apodus, *will accept freshly chopped uncooked red meat with a vitamin supplement, in place of their normal diet.*

**Fruit flies** (*Drosophila melanogaster*) - small insects barely 2mm(0.08in) long - are available as starter cultures from specialist suppliers. You can buy either the winged variety or, ideally, the recessive vestigial-winged strain, which are easier for both you and the lizard to handle. Alternatively, you can catch your own fruit flies by leaving a piece of rotting apple in a small jar outdoors in warm weather. Being so small, they are an ideal rearing food for newly hatched specimens or small lizard species.

Fruit flies are easy to culture in large quantities. Although commercial food preparations exist for fruit flies, it is cheaper to prepare your own. Simply place some finely chopped apple or banana mixed with bran to a depth of 2.5-5cm(1-2in) into a glass jar and insert a piece of wood so that it protrudes above the surface of the mixture. This will prevent the fruit flies drowning and also provides a platform on which they can mate and lay their eggs. Cover the top of the jar with cotton or muslin and secure this with a rubber band. Place the jar in a warm environment, about 20-25.5⁰C(68-78⁰F) is ideal. Lower temperatures will retard development and breeding capabilities, and too high a temperature will induce the growth of fungi, bacteria and small parasitic mites. To ensure a continuous cycle of fruit flies, simply prepare a new container every 10 days and introduce a dozen or so flies from the previous culture.

**Flies** (house flies, greenbottles and bluebottles or blowflies) are a valuable and nutritious food source for many lizards. In their larval form - as maggots, 'pinkies' and 'gentles' - they are readily and cheaply available from angling shops. Avoid dyed maggots, as these are known to contain harmful toxins. Place the maggots in a well-ventilated container filled with sawdust. If they are kept warm, they will pupate and hatch into flies in a very short time. Keep newly hatched flies in muslin cages containing a sweet solution, such as honey mixed with a multivitamin solution, which will increase their nutritional value. Make a small hole in the cage and plug it with a cotton bung. This will enable you to release a small number of flies into a jar, which you can then stand in the freezer compartment of a fridge for five minutes or so until the flies are sluggish. This makes them easy for the lizard to catch. As maggots are so cheap to buy, breeding your own flies is not worthwhile; it can be time-consuming and unhygienic, especially if the flies escape.

**Earwigs, woodlice, spiders and beetles** can be found in abundance under rocks nearly all year round (except in periods of frost and snow) and are an excellent way of varying your lizards' diet. Keepers sometimes raise the question of hygiene; although lizards do not seem to be affected by the bacteria accompanying such invertebrates, it is safer

to check the insects for any obvious external parasites before feeding them to the lizards. Avoid catching insects in areas that have been treated with a weedkiller or pesticide, as this could prove fatal.

Try sweeping a fine mesh fishing net through a grass meadow to catch winged insects both large and small. Greenfly are abundant in the warmer months and form an excellent food for small lizards. Caterpillars, slugs and snails will abound in vegetable plots and earthworms can be found in damp soil. Avoid the red-ringed earthworms found in compost heaps as they are poisonous.

**Feeding carnivorous lizards**
Larger lizards may need more substantial meals than just insects. You can offer them small rodents, chicks, fish or even tinned dog food, depending on the species. All these items can be frozen for future use, which is especially important in winter, when other foods may be absent. Make sure that you defrost the food completely - it should feel warm - as partly frozen or even cold food is harmful to lizards.

**Rodents** are suitable for large geckos, skinks, tegus and monitors; the size of the rodent depends on the size of the lizard. You can breed mice and rats at home or buy frozen ones. They are available in small or large quantities in a variety of different sizes: pinkies (newborn), juveniles, subadult and adult. Do not offer rodents captured from the wild, as they may harbour viruses or parasites dangerous to you and your lizard.

Always feed dead specimens to the lizard, as live rodents can inflict serious injuries. To kill a rodent quickly and cleanly, deliver a sharp blow to its head. Lizards that will react only to moving prey can usually be induced to feed if you wave the dead rodent about, clasped in a pair of long forceps or on a stick.

**Birds and eggs** are another nutritious food source for carnivorous lizards. Day-old chicks can be obtained cheaply in large numbers and are relished by larger lizards, such as tegus and monitors. Eggs are widely available and a valuable source of protein for larger skinks, iguanas, tegus and monitors.

**Fish** - whitebait, for example - is a highly beneficial food source for some semi-aquatic lizards, such as basilisks (*Basiliscus* sp.), water dragons (*Physignathus*), and tegus. Make sure it is completely fresh and offer it complete with bones, scales and entrails.

**Fresh minced meat, liver and canned dog meat** are relished by some lizards, including blue-tongued skinks (*Tiliqua* sp.), and other lizard species can be trained to accept such items. These foods are ideal during periods when other food is not available, but do not offer raw meat as the sole food, as it tends to be low in calcium and vitamin A.

## Feeding herbivorous lizards

Herbivorous, or plant-eating, lizards also need a well-balanced diet that will provide an adequate supply of vitamins and minerals, along with plenty of fibre to aid digestion. Fruit and vegetables are obtainable throughout the year, but you must ensure that they are fresh and have not been excessively treated with commercial fertilizers or pesticides.

Iguanas will eat fresh lettuce and cabbage leaves, but always offer them a wider variety of foods. These might include carrots (which yield vitamin A), chopped apple, pears and nectarines (all high in vitamin C), broccoli and spinach (rich in Vitamin B), along with weeds, grasses and hay (thoroughly washed of all pesticides) to provide extra roughage. You can add some crickets or crushed beetles, to the mixture to provide extra calcium and essential protein.

## Feeding omnivorous lizards

Many lizards are either occasionally or entirely omnivorous. Species of *Podarcis* and *Lacerta* are mainly insectivorous, but enjoy a small amount of ripe fruit, small weeds and honey in their diet. Insectivorous day geckos (*Phelsuma*) lick nectar in the wild and enjoy sugary substances, such as jam or honey, in captivity. Tegus and larger skinks are genuine omnivores, taking anything remotely edible.

The Australian blue-tongued skink, *Tiliqua scincoides*, appreciates banana in captivity - a food that is absent in its natural habitat!

## Vitamin and mineral supplements

As it is not always possible to offer a completely balanced diet in captivity, it is important to provide lizards with extra vitamins, minerals and trace elements to ensure their continued health and breeding potential. A wide range of suitable products is available in liquid, powder or tablet form. Some, mainly those in powdered form, are

Above: Phelsuma quadriocellatus *and other geckos store* calcium in lymphatic glands located just behind the head.

## Guide to essential vitamins

| Vitamin | Present in | Function |
| --- | --- | --- |
| A | Liver, cod-liver oil, green vegetables and carrots | Encourages good vision and healthy skin and glands. Strengthens immune system and aids growth and fertility. |
| B | Red meats, liver | Promotes cell division and nerve cell metabolism, as well as influencing the synthesis of protein. |
| C | Fruit and vegetables | Increases immunity to contagious diseases. |
| $D_3$ | Egg yolk, liver | Promotes healthy skeletal growth and repair. |
| E | Meat | Induces trouble-free livebearing or egg deposition and promotes muscular development. |
| K | Animal fats/oils | Increases the clotting ability of blood. |

formulated specifically for herptiles; ask your veterinarian or pet shop for details.

To administer a supplement, simply dust insects and invertebrates with some powder or add it to raw meat or chopped vegetables at alternate meals. Add multivitamin solutions to drinking water every time it is changed. Some species, such as day geckos, will benefit from a solution of water and vitamin supplement (4 parts water to 1 part supplement) and a little honey (1 part water/vitamin solution to 2 parts of honey).

Some lizards, including female geckos that lay hard, calcareous eggs, will eat crushed eggshell or powdered cuttlefish bone and store it in special lymphatic glands. This is later synthesized to assist in the development of strong-shelled, calcareous eggs. Provide supplies as required.

Above: *An agama drinks from a rainwater pool. In captivity, change the water daily, adding a vitamin supplement.*

### Frequency of feeding

The frequency of feeding depends on the age and type of species. Juveniles of most species require smaller, but more frequent, meals and the same can be said for naturally smaller, more active lizards, such as lacertids, agamids and smaller geckos. One feed a day should be sufficient for these lizards, although they will suffer no harm if the occasional day is missed. Herbivorous lizards need fresh food daily, as their food contains a higher proportion of water to solid matter.

As a general guide, a lizard needs enough food to keep it looking healthy, alert and active. Obesity due to overfeeding may lead to heart disease and a reduced reproductive response. You can allow the lizard's weight to increase slightly before hibernation or aestivation, as it will need enough fat reserves to last it through this critical period.

Problems with obesity are most likely to occur in slow, docile lizards and their diet should be strictly controlled. Feed them one large or several smaller food items on a weekly basis. Watch out for fighting or even cannibalism in some species.

### Water

Water is essential, both for the lizards to drink and to create a humid atmosphere, which is important for sloughing (see page 61) and also for any live plants in the vivarium. In most cases, you can provide water in a small dish, but lizards that enjoy bathing will require a larger container. It should be shallow, easy for the lizard to climb in and out of, and solid, so that it does not tip over, causing injury. Tap water is safe and there is no need to remove chlorine or reduce alkalinity.

Certain lizards, such as chameleons and geckos, will only drink droplets of water created by dew or rain. It is a good idea to finely mist the vivarium in the morning and evening, adding a multivitamin solution to the spray bottle.

### Force feeding

You should only consider force feeding a lizard when it has refused food for an extended period (see page 40). Use a smooth, flat plastic utensil - not a metal one - taking care not to damage the lizard's jaw or teeth. Always use sterilized equipment and make sure your hands are clean. Seek help if you are inexperienced.

If the lizard is small, hold it firmly in one hand and insert the lever gently into the corner of the mouth, which should encourage the lizard to open its jaws. You may need help with a larger lizard; one person to hold the lizard steady while the other opens the jaw and inserts the food. If the lizard has a flap of skin or dewlap on the lower jaw, you can open the mouth by gently pulling this. Lubricate small pieces of food with water or a multivitamin solution, so that they slide easily down the throat. Use a pair of forceps to assist the food into the gullet opening.

# BREEDING LIZARDS

Reproduction and courtship are fascinating and important aspects of keeping lizards in captivity. They are also standards by which you can judge just how successfully you are maintaining your lizards. It is safe to assume that if breeding does occur, then the environment, conditions and food you have provided are virtually perfect.

Lizards fall into three reproductive groups, in ascending evolutionary order. In the first group are the oviparous species that deposit eggs. Group two contains the ovoviviparous species that retain their eggs within the body. The embryo takes nourishment from a yolk-sac and depends on the mother only for physical protection. In the final group are the viviparous species. Here, the embryo develops inside the mother, taking nourishment from her, and is born alive at an advanced stage of development.

Some species are more difficult to breed than others. They may require particular stimuli or certain environmental conditions that are not easy to provide in captivity. But other lizards are very easy to breed and will mate almost immediately. If you are a beginner, it is clearly a good idea to start by breeding the 'easier' species. Experience will give you the confidence to try the more 'difficult' species later on.

## Sexing lizards

In order to breed lizards, you must first obtain a sexed pair, i.e. a male and a female. Generally speaking, specialist reptile dealers will be able to determine the sex for you, but occasionally you may have to establish this for yourself.

The easiest lizards to sex are those that show clear sexual dimorphism - for example, a difference in size, appearance or coloration. Lacertids, some iguanids, agamids and geckos all fall into this category. Quite often, males are larger, show brighter colours and develop skin crests or horns on their head, legs, back or tail, whereas females are duller in coloration and not as well - or not at all - adorned. In some species, dimorphism may be apparent only during the breeding season, while in others it is possible to determine the sex of a species once it is adult, as juveniles of both sexes are brightly coloured. Where no obvious gender differences exist, behaviour may be a useful guide, especially during courtship, when males may fight or perform courtship displays.

The sex of certain geckos, iguanas and agamas can be established by studying the cloacal and lower ventral region. In male geckos, anal or preanal pores are present, and male agamas or iguanids have thigh or

Above: *The underside of a male wall lizard,* Podarcis pityusensis *(left) showing the* *enlarged scales, or femoral pores. These* *are often absent in females (shown right).*

femoral pores. These pores are also present in the females of some species, but are rarely as well defined. Sex can also be determined by measuring skull sizes or tail base diameter and length, but you need an experienced eye to make a reliable judgement.

In species where there are no visible differences between the sexes, it is possible to carry out an internal examination using one of the solid probes recommended for reptiles. A well-lubricated probe is inserted into the cloaca and gently pushed towards the tail end. A male has an inverted hemipenis - an inside-out double penis - which extends internally along the base of the tail. Consequently, the probe will travel further inside the male than the female. This examination *must* be executed with the utmost care and gentleness, otherwise

Above: *A pair of geckos,* Gonatodes vittatus. *The brightly coloured male is shown above, the drabber female below.*

permanent internal damage and even death may result. If you are inexperienced or unsure, ask an experienced reptile handler to sex your lizard.

### Conditioning and stimulation

In order to breed, lizards must be healthy, well fed and appear ready to mate. The first two prerequisites are essential for those lizards that hibernate or aestivate and, generally speaking, most lizards will not breed if they are out of condition.

Owning a sexed pair is no guarantee that breeding will automatically follow. Usually, the sexual instinct of the lizard must first be aroused by one of several stimuli, the most common being a change in climatic conditions. This is indeed crucial for temperate species, where breeding is triggered by an increase in temperature and day length. As we have seen (page 16), some species undergo a period of torpidity during cold winter months and this must be simulated in captivity. The easiest method is to place the *healthy* specimens into an escape-proof aquarium filled either with dry straw or shredded newspaper. Place the aquarium and its occupants into a cold but frost-free position, say a garage or cellar, at a minimum temperature of 4-7⁰C(39-44⁰F) for about six to eight weeks. On warming up - and with an increased supply of food - the male's testosterone level increases significantly and egg development begins in the female.

Tropical species can also benefit from a controlled change in temperature. They will tolerate a drop of about 2-5⁰C(5-10⁰F) over a period of a few weeks. When the temperature suddenly increases, courtship and mating should begin. This method works well with tropical geckos, such as the leopard gecko (*Eublepharus macularis*) and is worth trying with other species. Species from savanna or semi-desert regions may aestivate during the drier, hotter parts of the year and the advent of rain or lower temperatures often stimulate breeding. Since an increase in food, rainfall or humidity can all stimulate breeding, knowing something about your lizards' natural environment will help you to identify the conditions required to breed them.

Other stimulants include pheromones - natural body scents given off by females when they slough their skin or are sexually active. These often drive males into a frenzy of fighting and courtship displays.

Breeding can be discouraged by overcrowding, lack of food or obesity, 'bullying' by different, larger species or by larger individuals of the same species. Old age, illness and disease or incorrect sex ratios can all have an adverse effect.

**Sex ratios**
In wild populations of certain species - day geckos (*Phelsuma* sp.) or typical lizards (*Lacerta* and *Podarcis* spp.), for example - an excess of males will often compete for a single female. In captivity, this behaviour can be simulated, but only when males come into breeding form. At other times, the consequences could be highly dangerous, as males will fight for territories or dominance outside the breeding period, causing disfiguring injury and death. In a captive environment, a viable ratio would be in the region of three males to one female. In leopard geckos (*Eublepharus*) and certain agamids, the situation is completely reversed. These are polygamous species, where a single male will mate with a number of females. In such cases, you can keep two to seven females with one male. For other species, an equal ratio or one male *only* to any number of females is advised.

Left: *The males of some lizard species are highly territorial. For example, be sure to keep male leopard geckos apart; if not, they will fight and risk sustaining very serious injuries.*

Right: *During the courtship period, male diurnal lizards often become brightly coloured to attract females. Pictured here are male tree agamas,* Calotes versicolor.

## Mating and frequency of breeding

Courtship behaviour is described on page 17. The frequency with which lizards breed depends on how long they remain active throughout the year. For example, lizards from temperate regions, such as many of the lacertids, some anguids and skinks mate and breed only once a year, as do some of the lizards from desert regions. In these cases, breeding frequency is mainly governed by the availability of food and the growth or survival factor of newly born lizards. If lizards were born just before the onset of winter or drought they would not stand much chance of acquiring enough body fat to sustain them during hibernation or aestivation. For the same reason, many of these lizards are livebearers, and their offspring therefore have an improved chance of survival. In contrast, lizards from tropical or equatorial regions may breed several times throughout the year, since they enjoy consistently higher temperatures, ample rainfall and plentiful supplies of food.

## Gravidity

If mating is successful, the female will begin a period of gravidity (pregnancy). In some species, signs of gravidity are easily observed, whereas in others the condition is difficult to determine. In geckos, simply allow the specimen to climb onto a sheet of glass and shine a powerful light on its back. By looking under the glass, you can see whether any eggs are developing. With other species, gently allow a female lizard to move through your hand, with your fingers on her back and thumb on the underside. If she is gravid, you will be able to feel slight swellings. There may also be a change in the female's behaviour; for instance, she will be agitated by other lizards round her and, in the early stages of gravidity, will eat more frequently. Egg layers will eat calcified substances. It is a good idea to separate the female during gravidity and keep her in warm, humid conditions, giving her plenty of multivitamin 'dressings' on a varied diet and changing her drinking water frequently.

The gestation period varies with each species, but is longest in the viviparous species. When a female is due to deposit her eggs or young, be sure to provide a container lined with moist, sphagnum moss or sand.

Once the eggs are deposited, the female will show no further interest in them. The

Below: *A pair of Iguana iguana mating. Healthy lizards often breed in captivity if the conditions are right.*

Below: *Incubating lizard eggs is not difficult if a few simple rules are observed. The soft, leathery eggs produced by most lizards need a clean, humid atmosphere in order to develop. Geckos, however, lay brittle calcareous eggs.* *Place these on dry paper towelling and give them a fine misting from time to time. Temperature can play an important part in determining the sex of the hatchlings of species such as leopard and day geckos. (See also page 54).*

### Guide to egg incubation periods

| Species | Incubation period at 29°C(84°F) |
|---|---|
| Moorish gecko (*Tarentola mauritanica*) | 4-12 days |
| Blue-tailed skink (*Eumeces skiltonianus*) | 25-35 days |
| House gecko (*Hemidactylus frenatus*) | 35-45 days |
| Green lizard (*Lacerta viridis*) | 40-50 days |
| Horned lizard (*Phrynosoma* spp.) | 50-65 days |
| Green iguana (*Iguana iguana*) | 60-75 days |
| European agama (*Agama stellio*) | 68-85 days |
| Tokay gecko (*Gekko gecko*) | 120-200 days |

exception to this general rule are some North American skinks (*Eumeces*), which wrap their body and tail around the small clutch until it hatches. Lizards lay two types of egg; gecko eggs are hard, brittle and calcareous, but most egg layers produce softer, pliable parchmentlike eggs. The same approach to incubation will suit both types.

### Incubating eggs

Always try to remove eggs from the vivarium a few hours after they have been deposited. Having removed the eggs, it is essential to retain them in the same position, i.e. do not turn them, otherwise the yolk may actually envelop and suffocate the fixed embryo. The best method is to mark the top of the egg gently with a waterproof marker pen. Lizard eggs do not require expensive incubating equipment, but they do need a warm, sterile and humid atmosphere. The best way to incubate them is in a sandwich box, half-filled with vermiculite or

horticultural perlite. Run tap water into the container and then drain away any excess water so that the vermiculate or perlite chips are uniformly moist but not soaking wet. Settle the egg into this medium so that about half of it is visible, and firmly secure the lid onto the container. Place the box in an airing cupboard or a room heated to about 24-32°C(75-90°F), but never in direct sunlight. Do not allow the temperature to fall below this level for extended periods. The warmth will create an ideal, humid atmosphere inside the box and the moisture contained in the vermiculite should be sufficient to last throughout the incubation period. Should the eggs begin to shrivel, mist them gently with tepid water. Discoloring and fungal attacks on the eggs are quite natural, because the atmosphere within the box is ideal for encouraging the growth of microorganisms. Never discard such eggs, however, as perfectly healthy lizards will inevitably hatch out from them.

Some gecko eggs adhere so firmly to the side of the vivarium or onto a branch that removal is impossible without breaking the fragile shell. If the eggs cannot be removed, then provide some sort of protective cover, such as a transparent beaker or plastic bag, and lay a moist paper towel below the egg to maintain humidity.

### Determining sex by temperature

The incubation period of eggs varies according to species and temperature. In some species, it may be possible to control the sex of the eggs by incubating them at certain temperatures. At higher temperatures, say 29-33°C(84-92°F), all males are produced, whereas at 23-27°C(74-80°F), all females. Between 27 and 29°C(80-84°F) both sexes develop. The critical period for this determination seems to be during the latter part of incubation, when the sexual organs are developing. Fluctuating temperatures throughout the whole range on a daily basis will also produce both sexes.

Below: *This series of photographs illustrates the stages in the hatching of the eggs of the day gecko Phelsuma laticauda. Four calcareous eggs are shown adhering to a branch and here the embryos develop until one begins to break the* *shell with its eggtooth, six to eight weeks after being laid. Having absorbed its yolk sac, the hatchling emerges, a process that may take several days. The miniature replica of the adult then sloughs its skin, before hunting for its first meal.*

### Guide to gestation periods in livebearing lizards

| Species | Gestation period |
| --- | --- |
| Common lizard (*Lacerta vivipara*) | 90-110 days |
| Desert night lizard (*Xantusia vigilis*) | 100-120 days |
| Slow-worm (*Anguis fragilis*) | 110-140 days |
| Blue-tongued skink (*Tiliqua scincoides*) | 90-150 days |

### Hatching and rearing

If an egg has not hatched after about 200 days, you can assume that it is either infertile or that the embryo has died during the incubation period.

Hatching lizards are equipped with a sharp toothlike projection (an 'eggtooth') on the snout, with which they cut through the relatively tough eggshell. Hatching may take several days and, on leaving the empty shell, the miniature lizard will shed its skin along with the eggtooth. At this point, remove the juvenile to a rearing container provided with ample heat, humidity, food and water. Empty calcareous eggshells can be fed to the adult female as a valuable source of calcium, but

discard empty parchmentlike eggshells to reduce the risk of infection to other eggs.

Live-born or hatchling lizards need the same care for the first few weeks of their life. They are rather delicate creatures, requiring small foods, such as greenfly, fruitflies or small, soft grubs and caterpillars, all liberally dusted in multivitamin powder. Keep each lizard on its own, so that none need compete for your attention. If well fed, they will grow very quickly; some geckos can reach maturity

Above: *Not all lizards lay eggs. The slow worm, for example, gives birth to live young, enclosed in a thin membrane. They have a better chance of survival than eggs, which need warmer conditions to incubate.*

within six months, while slow worms may take eight to ten years. Hardy lizards should be overwintered in warmer conditions for the first year, after which they can be treated in the same way as adults.

# HEALTH CARE

Lizards are highly resistant to disease and infection, but even the healthiest specimens can occasionally succumb to illness. In most cases, disease is already present in newly acquired specimens, particularly those from the wild. Therefore, it is essential to place all new acquisitions in quarantine for six to ten weeks before adding them to an established collection. During this time, you will be able to assess whether the lizard is showing any signs of ill-health and can take the appropriate measures. Problems can also be caused by feeding lizards with cold, infected, old or unsuitable food, or by inadequate vivarium cleaning and maintenance.

Observe all lizards carefully to ascertain that they are behaving and feeding normally. From time to time, examine them more closely for any signs of parasitic, viral or other physical disorders. If a single specimen is ailing, isolate it quickly and keep an eye on the other specimens. Check female lizards occasionally for signs of gravidity.

Fortunately, most disorders are easily cured if treated promptly, but treating smaller lizards can be very difficult. For example, you may have to consult a veterinarian if surgery is needed to remove retained eggs. Veterinarians are not always familiar with herptile ailments, but their experience with other animals will enable them to demonstrate treatment procedures and explain how to administer medicines. They may even refer you to someone with specialized experience. Joining a herptile society will give you access to a valuable source of help and advice.

**Parasites**
Parasites are broadly divided into two types: external (ectoparasites) - mainly mites and ticks - and internal (endoparasites), such as nematodes and tapeworms.

**Mites** (*Ophionyssus*) are dark, pinhead- sized parasites that feed on the blood of their host. They are often present in huge numbers, mainly congregating on softer areas, such as under the scales, and are particularly common around the eye and the eardrum. They are quite inconspicuous unless you examine the lizard closely.

The most effective and familiar way of eradicating mites is to place a small dichlorvos strip in the top corner of the vivarium. Encase it in a wire gauze or muslin 'cage' so that the lizard cannot touch it and leave it in position for 36 hours. Repeat this procedure every ten days until the problem is eliminated. Never leave a dichlorvos strip in

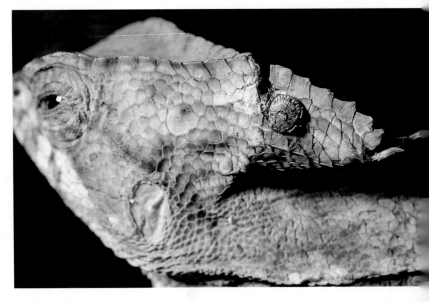

Right: *It is quite easy to detect the presence of ticks* (Ixodidae*) on lizards. They are clearly visible, usually attaching themselves to softer regions of skin, particularly between scales, in armpits and leg joints or around the ear opening. If left untreated, ticks can grow surprisingly large and will quickly spread, causing wounds, transmitting blood diseases and subjecting the reptile to a great deal of stress.*

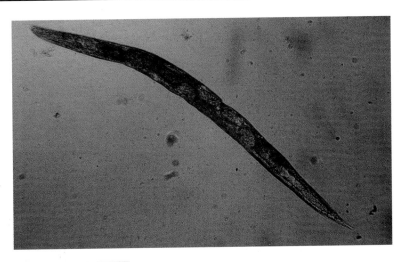

Right: *The presence of intestinal worms, such as the nematode larvae, can be detected by examining the lizard's faeces. Improve general standards of hygiene.*

Below: *Tapeworms are sometimes present in lizards and can have serious consequences if left unchecked. They are easy to treat, and a veterinarian can advise on effective remedies.*

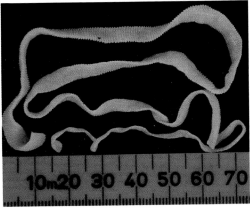

the vivarium for longer than necessary, as it can have dangerous side effects on lizards, particularly in a small vivarium. If the infestation continues, bathe the lizard for at least an hour in a 0.2 percent solution of a proprietary remedy containing dichlorvos and trichlorvos, available from chemists. For heavier infestations, a 0.4 percent solution may be necessary, but do not make the solution too strong; this is a particularly potent chemical, so take care.

Bear in mind that some remedies are toxic to members of the family Gekkonidae and Lacertidae and to other lizards with a thin epidermis. You could try a combination of the dichlorvos strip and dabbing the mites with an alcohol solution, such as methylated spirits. Thoroughly disinfect the vivarium and its decorations or discard the decorations. Repeat this treatment every five to seven days until the mites are eradicated. If they persist, consult a veterinarian.

**Ticks** (Ixodidae) are larger, more visible parasites, often attached between the lizards' scales and most commonly associated with wild-caught specimens of larger lizards. Do not pull ticks directly away from the skin, as their mouthparts will invariably be left behind to cause infection. If the infestation is light, bathe the lizard in a proprietary remedy as described for mites and at the same strength solution. For heavier infestation, wipe the lizard down with alcohol. In both cases, use a dichlorvos strip in the vivarium. Remove the dead ticks with forceps, carefully turning them 'head-over-heels', thus unhooking the mouthparts.

If left unchecked, all ectoparasites will stress and irritate the lizard, due to the constant scratching. Anaemia and death eventually follow. Equally dangerous is their ability to carry and transmit diseases or blood parasites from one lizard to another.

**Blood parasites** include the single-celled *Entamoeba invadens*, which is both dangerous and highly contagious in lizards, especially where hygiene is poor. Symptoms include constant regurgitation of food and depositing pungent, slimy faeces. A post-mortem often reveals complications of the liver and large intestine. A veterinarian will prescribe an appropriate drug and daily doses of multi-vitamin solution, administered by syringe.

**Intestinal worms**, including various roundworms and tapeworms, and flukes (trematodes) can enter and attack lizards, usually through their food or unclean water or via another infected lizard or its faeces. White threadlike roundworms are found in concentrated numbers in the faeces. Some are only visible under a microscope, so it is a wise precaution for a veterinarian to examine the faeces of all your lizards at least annually, or monthly for specimens in quarantine. This will reveal whether endoparasites (or even *Salmonella* bacteria) are present and enable you to take swift remedial action. Roundworms cause emaciation and damage to the intestinal wall, resulting in a poor appetite. Proprietary dog and cat worming preparations are too toxic for all but the larger lizards, so the veterinarian may prescribe mebendazole or a similar drug.

Tapeworms are long and flat and segments, or egg cases, can often be seen in the faeces. Heavy infestations are particularly hazardous, as they will cause severe malnutrition, and tapeworm hooks, spines and suckers may damage or infect the intestinal wall. Infested lizards should be treated with praziquantel or bunamidine, as prescribed by a veterinarian.

**Lung mite, or pentastomid worm**, is particularly common in lizards imported from Southeast Asia. As you might expect, it is found mainly in the lungs or under the skin. Take particular care with infected lizards, as the parasite is transmissible to man. Treatment is very effective with dichlorvos (see page 56).

**Bacterial diseases**
In the wild, many bacterial pathogens occur in lizards but, because they are found in such small quantities, they rarely cause any problems. Unfortunately, the stress and damage caused during transporting lizards can result in the rapid proliferation of these bacteria, with very dangerous consequences. Swift medication is needed to prevent the bacteria multiplying and strict hygiene procedures are essential. Neglecting routine hygiene can lead to bacterial infection, even in the healthiest captive lizard.

**Epidermal necrosis** caused by the bacterium *Pseudomonas* affects the skin of lizards, resulting in abscesses and inflammation. Poor hygiene, poor quality food, an unbalanced diet or insufficient UV lighting

are usually to blame. Treat the condition swiftly, preferably consulting a veterinarian, who will lance any abscesses and apply a weak hydrogen peroxide solution or administer an antibiotic.

**Mouthrot** (jaw suppuration) is mainly caused by the same bacterium, *Pseudomonas*, that causes epidermal necrosis. Under optimum conditions, it can attack the soft mouthparts of lizards, especially if there is some previous damage. The infection causes excessive cheesy discharges, foul odours, loss of teeth and inflammation, with the jaws constantly gaping. Eventually, it will spread to attack the digestive tract. In its early stages, mouthrot can be disinfected with a 2 percent hydrogen peroxide solution and brushed with a full strength iodophor, a volatile compound of iodine that acts as an antiseptic when carefully applied. Advanced infections may need antibiotic treatment, and are best referred to a veterinarian. An increased supply of multivitamins, either in food or in a solution syringed down the lizard's throat, is beneficial.

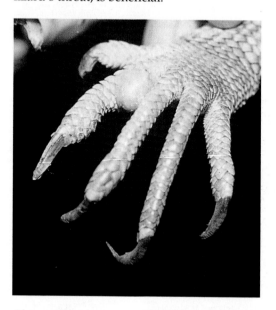

Above: *This common iguana has an abscess on the foot. It may be the result of poor vivarium hygiene and bacterial infection of a small lesion.*

Right: *Warts or viral papillomas on a green lizard. They are spread by direct contact with other infected lizards. The virus enters via lesions or sores.*

**Pneumonia** can afflict many lizards, particularly tropical species, which are susceptible to bacteria attacking the lung tissue. It can be caused by permanently overheating or underheating the vivarium, poor ventilation or severe stress. Symptoms range from constant wheezing and difficulty in breathing to excessive mucal discharge from the mouth and foaming from the nostrils. Apart from providing the correct heating levels (including lower night-time temperatures) and adequate ventilation, a course of antibiotics administered via the drinking water or intramuscular injections may be needed.

**Salmonella** bacteria can infect lizards, but the majority of strains rarely cause illness. A severe infection of a dangerous *Salmonella* bacterium will cause regurgitation of food and slimy faeces. It can cause death and may be transmitted to humans. By using fresh, uncontaminated food, sterilizing equipment and tools, and implementing strict hygienic conditions it is possible to reduce the risk of *Salmonella* infection to a minimum.

**Nutritional disorders**

These common disorders are most likely to occur when the diet is inadequate or unbalanced. An insufficient food supply can be a problem during certain periods of the year and it is essential to find suitable substitutes or an alternative source of supply. Imbalances in vitamins or minerals can have unpleasant consequences, but so many multivitamin preparations are available today that there should be no problem. Some vitamins, particularly vitamin $D_3$, may need synthesis under the influence of ultraviolet irradiation. Here we look at the more common vitamin and mineral deficiencies associated with lizards.

**Vitamin A deficiency** occasionally occurs in lizards. A severe deficiency can cause loss of appetite, swelling of and around the eyes, night blindness and excess mucus in the eye. (The eyes of a healthy lizard are protected by a watery eyewash. In cases of vitamin A deficiency, a greater proportion of mucus is produced in the eyewash, and this sticky substance can harden on the outer eye and

impair vision. Symptoms include streaming eyes, possibly with lumps of jelly-like substances over them.) Vitamin A deficiency can also affect skin development, making it weak, brittle or soft. Cuts or lesions may appear, sloughing will be difficult and external growths can be clearly seen. Any weakness in the skin will leave the lizard more open to infection. Adding a few drops of cod liver oil to the normal diet will help, but lizards in an advanced state of deficiency may need injections of vitamin A.

**Vitamin B, riboflavin deficiency** occurs mainly in iguanids, agamids and some lacertids. It can be difficult to identify positively in its early stages, as symptoms appear identical to natural lizard behaviour. These include tail and limb vibrations, otherwise associated with excitement, disturbance or courtship. In later stages, the deficiency is lethal, with the lizard refusing food, flipping onto its back and convulsing before becoming paralyzed. Vitamin $B_2$ deficiency can be averted by increasing the multivitamin supplement in food, but when it reaches an advanced stage, a veterinarian may need to administer high-dose, intramuscular injections of thiamine-type drugs until the symptoms cease.

**Vitamin D and calcium deficiency**. In many lizards, particularly juveniles and females, insufficient calcium in the diet will seriously retard skeletal and egg development. Low calcium foods, such as mealworms, vegetables and pure meats, as well as foods containing a high proportion of phosphorous, will cause an imbalance. Skeletal and spinal deformities result, eggs are produced with very fragile shells, muscle contraction is limited and the ability of blood to clot is reduced. Even where sufficient calcium is given, it is usually in the form of insoluble salts that must be converted into smaller, easily absorbed ions. In lizards, the vitamin $D_3$ (not $D_2$ as in mammals) is essential for this conversion process and is produced in the stomach under the influence of ultraviolet radiation. In captivity, UV or similar tubes (see page 28) can be used to good effect, although they cannot be relied upon to provide all the necessary vitamin $D_3$. Combined calcium and vitamin $D_3$

preparations are available commercially or in multivitamin supplements. Extra calcium can also be provided in the form of ground eggshell or cuttlefish bone. Pure vitamin $D_3$ is not recommended, as excessive doses can have serious side-effects; too much calcium released into the bloodstream eventually accumulates and causes heart and circulatory complications.

Vitamin $D_3$ deficiency can cause a number of other disorders, such as decreased egg production, egg or embryo abortion, deformed embryo development and osteoporosis (weakened bones), all of which can be avoided by increasing supplemented vitamin $D_3$. Other problems, such as egg retention, may require more specific treatment (see below), but paralysis and skeletal deformity are usually permanent.

**Egg retention**
Stress or lack of sufficient vitamin $D_3$ can cause a female to retain her eggs; alternatively, egg-binding can make it

Below: Lacerta viridis *with a swollen eye. Such injuries* *may be caused by fighting, bullying or even overhandling.*

Above: *A leopard gecko sloughing and devouring its old skin.*

*A humid atmosphere is vital for lizards to slough successfully.*

impossible for her to deposit them. This retention leads to egg decay in the womb and the eventual death of the lizard. A veterinarian can inject the lizard with a contraction-inducing drug, but if this fails, surgical removal is required, although this procedure is impossible on small lizards.

## Sloughing
As lizards grow, they need to shed their outer layer of skin to accommodate the extra body area. In lizards, unlike snakes, this is usually a messy and irritating affair, with the dead epidermis coming off in small shreds. For reasons of hygiene, always remove dead skin from the cage, although geckos will usually devour it anyway. The frequency of sloughing depends on the species, its age and growth rate, but a healthy young leopard gecko (*Eublepharus macularis*), for example, will shed its skin every six weeks.

Occasionally, a lizard may have difficulty in shedding its skin. The danger with sloughing problems is that the lizard quickly uses up its energy and becomes exhausted, highly irritated and stressed. The presence of ectoparasites or a physical disorder may be the cause, but problems are usually related to incorrect climatic conditions. Failure to provide sufficient humidity makes the old

skin dry so that it adheres to the new skin. Sloughing problems can also be the result of giving too little or too much vitamin supplement, so it is worth considering an increase or decrease in these supplements. Bathing the lizard in a slightly saline solution until the old skin is soft will make it easy to remove with tweezers.

## Poor appetite
All the diseases and ailments described so far may result in a loss of appetite, but stress, poor hygiene or an incorrect diet are other possible explanations. Occasionally, a new acquisition may refuse food until it has settled down. In any case, it is certain to be stressed by transportation and its new surroundings. Bullying or dominance within various groups of lizards can generate stress, usually in juvenile, immature or elderly individuals. In this case, separation is the only answer, unless you can provide a more extensive set-up. Inadequate vivarium conditions, which can include the structure of the vivarium itself - the openness of an all-glass vivarium may be unsettling - but may also stem from temperature, humidity, lighting, decoration or even continual disturbance by the keeper, can all cause stress and loss of appetite. Poor hygiene is bound to cause stress and illness, so regular and thorough general maintenance is essential. Clearly, a balanced and adequate diet also has a vital role to play.

61

# Lizards • Species Section

Many lizards will live quite happily in a relatively small vivarium, as long as they have sufficient space to move around freely. Large species, such as monitors and typical iguanas, obviously require more spacious accommodation, but this also applies to wall lizards, agamids and anoles if you wish to house them in groups. As a general rule, the length or height of the vivarium (depending on the terrestrial or arboreal habits of the lizards) should be approximately five times the total length of each of the lizards to be housed in it, up to a maximum of 12 individuals. Therefore, a suitably proportioned vivarium measuring 120x45x45cm(48x18x18in) could accommodate up to 12 wall lizards, each 23cm(9in) in length.

You may wish to keep several species of lizard together in one of these larger vivariums. This is possible, as long as you take into account certain factors, such as climatic and habitat requirements and male (or sometimes female) territorial behaviour. For example, *Mabuya*, a diurnal, ground-dwelling skink, wall lizards or agamids, and a nocturnal gecko could live quite peacefully in the same vivarium. The wall lizards or agamids will spend most of their time on top of rocks and logs, the skink will occupy the base area and the gecko will be active at night when the other lizards are resting. Nocturnal geckos can be housed with any similarly sized diurnal lizard species.

In the wild, several species of, say, wall lizard that might normally be expected to fight, are often found living quite close together. To avoid competition, each occupies a particular area within the same vicinity, be it a shady or sunny spot or an area near the water. This behaviour is also seen in captivity.

Of course, some lizards are simply not compatible with other species or even with other members of their own species. The Tokay gecko (*Gekko gecko*), different species of day gecko (*Phelsuma*) and the leopard gecko (*Eublepharus*), particularly male individuals, will not tolerate other species. Adult blue-tongued skinks (*Tiliqua*) and certain chameleons will fight and inflict nasty injuries on one another. Individuals of these species are best kept apart, except during the breeding period.

Some lizards cannot adapt to life in captivity, while others, such as certain chameleons, are known to have a naturally short lifespan or to require specialized climatic conditions. However, this still leaves a wide variety of lizards from which to choose.

Below: *The unique
giant Solomon Islands
or monkey-tailed skink,
Corucia zebrata, has
a prehensile tail and
sharp claws suited  to
its arboreal lifestyle.*

## FAMILY: AGAMIDAE

Although the agamas form a large family, dealers do not seem to offer them extensively. This may be due to their requirements in captivity, namely a spacious vivarium, 'hot spots' and plenty of food. Their social behaviour is very complicated, with aggressive males fighting for both territories and females, a factor which can make them difficult to keep. Many species grow over 30cm(12in) long and are better suited to zoological collections.

The majority of agamids lay 4-12 eggs, which hatch within 35-70 days. Rear the hatchlings in individual containers, offering them small crickets, grubs and mealworms.

### European agama
*Agama stellio*

**Distribution:** Mediterranean, North Africa.
**Length:** 24-30cm(9.5-12in).
**Sex differences:** Dominant males are more colourful, with well-developed femoral pores.
**Diet:** Particularly fond of ants and termites; will take small insects, flowers and soft fruit.
**Ideal conditions:** Large semi-desert vivarium with plenty of rocks and hiding places. Ideal temperature range 24-32°C(75-90°F). Can be kept in a greenhouse during warmer months.
**Hibernation:** No, but reduce temperature by 9°C(16°F) for two or three months in winter.
**In captivity:** These agamas need a spacious enclosure and are more suited to the experienced hobbyist.

**Other species of interest:** Other species are occasionally available. The Caucasian agama, *A. caucasica*, from western Asia, including Turkey, grows to the same size as its European counterpart, but has more striking markings. The common agama, *A. agama*, from Africa is a larger species, attaining 35cm(14in). Males have a red-tinged head in the courtship season and are particularly aggressive at this time.

Two species suitable for the smaller vivarium are the toad-headed agamas, *Phrynocephalus helioscopus* and *P. mystaceus*. Both inhabit dry regions of western and central Asia and rarely grow longer than 10-12cm(4-4.7in).

Above: *The thick horny skin of the European agama,* Agama stellio, *provides protection against the hot arid conditions that prevail in its natural habitat.*

Left: *The striking coloration of this male* Agama agama *is displayed when the lizard is excited or seeking to attract a potential mate in the breeding season.*

## FAMILY: ANGUIDAE

This family is made up of three subfamilies: the Anguinae, the Gerrhonotinae and the Diploglossinae. Species from the last family are rarely obtainable, but those from the first two are widely available and very popular.

SUBFAMILY: Anguinae

### Slow-worm
*Anguis fragilis*

**Distribution:** Throughout western Asia, North Africa and Europe, except Scandinavia and Ireland.
**Length:** 30-50cm(12-20in).
**Sex differences:** Females have darker sides, with a black vertebral stripe. Males are sometimes blue-speckled.
**Diet:** Earthworms, grubs and caterpillars, but especially fond of small white and grey slugs.
**Ideal conditions:** An aquarium lined with 10-15cm(4-6in) of loose substrate, such as damp sphagnum moss. Maximum temperature 24$^0$C(75$^0$F). A good species for the outdoor vivarium.
**Hibernation:** Yes.
**In captivity:** Very easy to maintain and long-lived. Male slow-worms fight fiercely during the breeding season. The female is live-bearing, giving birth to 6-24 young. Juveniles are a striking coppery colour and feed on very small slugs and earthworms.

Above: *The slow-worm,* Anguis fragilis, *is normally long-lived, but is often mistaken for a snake and killed as a result.*

**Other species of interest:** Apart from the subspecies, *Anguis f. fragilis*, there is a particularly attractive subspecies from eastern Europe and Turkey, *A.f. colchicus*, which has blue spots along its back.

SUBFAMILY: Gerrhonotinae

### Northern alligator lizard
*Gerrhonotus coeruleus*

**Distribution:** Mountainous areas of northwestern USA.
**Length:** 30cm(12in).
**Sex differences:** The sexes are very similar,

Below: *A pair of* Anguis fragilis colchicus *in a typical courtship embrace.*

Above: *In common* *species,* Gerrhonotus
*with other alligator* kingi, *prefers to live in*
*lizards, this Arizona* *cool moist conditions.*

but males have a somewhat thicker tail base.
**Diet:** Insects, slugs and small snails.
**Ideal conditions:** Plenty of branches and
rocks, frequently misted, as these lizards
prefer high humidity. Temperature range of
18-25°C(64-77°F).
**Hibernation:** Yes, at 7-12°C(45-55°F) for three
months.
**In captivity:** Docile and quite easy to
maintain and breed.

Unlike other anguids, alligator lizards have
well-developed limbs. The northern alligator
lizard is mainly olive-coloured with dark
flecks. It is livebearing, giving birth in late
summer to 4-10 young.

**Other species of interest:** The southern
alligator lizard, *G. multicarinatus*, comes from
the lowlands of southern Washington south
to Baja California It is slightly larger than the
northern species, reaching 35-40cm(14-16in)
and is oviparous, laying 4-13 eggs, which the
female protects. These hatch after 40-50 days.
The southern alligator lizard cannot be
hibernated, but needs a drop of around 5-
7°C(8-13°F) for a few months in winter.

### European glass lizard; Pallas's glass lizard; Scheltopusik
*Ophisaurus apodus*

**Distribution:** Eastern and southeastern
Europe to western Asia.
**Length:** 100-120cm(39-48in).
**Sex differences:** No visible difference. Use a
probe to determine sex.
**Diet:** Insects, slugs, snails, pinkie mice. May
accept strips of fresh red meat.
**Ideal conditions:** A large dry vivarium
containing rocks, logs and a newspaper base.
Mist the vivarium frequently and maintain
the temperature at 20-30°C(68-86°F).
**Hibernation:** No, but a drop to around
12°C(55°F) for several months is beneficial.
**In captivity:** Although shy and generally
docile, these powerful lizards can give a
painful bite. Long-lived.

**Other species of interest:** The American
glass lizards, *O. attenuatus* and *O. ventralis* are
occasionally available. Both are similar to -
but smaller than - the European species,
measuring up to 90cm(36in). The Chinese
variety, *O. harti*, is very attractive, with an
olive back and light blue sides. All require
the same treatment and breed once a year, the
female laying 6-12 eggs. These develop
quickly, with the 13cm(5in) hatchlings
appearing 20-30 days later.

Below: *The long tail of* *breaks easily when*
*the glass lizard,* *grasped, hence the*
Ophisaurus apodus, *common name.*

**FAMILY: ANNIELLIDAE**
This small family of Californian legless lizards is restricted to just two species. Their secretive habits make them difficult to obtain so, when they are available, they tend to be quite expensive. However, they are worth trying, their silvery, highly polished appearance making them attractive captives.

**Geronimo legless lizard**
*Anniella geronomensis*

**Distribution:** San Geronimo and lower California, USA.
**Length:** 10-18cm(4-7in).
**Sex differences:** The tail base is swollen in males, but probing may be necessary to determine sex.
**Diet:** Mainly insects and spiders.
**Ideal conditions:** Subterranean aquarium, lined with 13-20cm(5-8in) of damp, sterilized, loamy soil. Temperature range about 18-24°C(64-75°F).
**Hibernation:** No.
**In captivity:** Easy, but dislikes being kept in dry conditions.

**Other species of interest:** The Californian legless lizard, *A. pulchra*, is very similar in size and habits, but has a wider distribution along coastal California. Both species give birth to between one and five live young, which are creamy coloured and only 4cm(1.6in) long. They need close attention for the first few months.

Below: *The strange wormlike Anniella geronomensis has smooth scales that allow it to move easily through the moist sand dunes and soil of its natural habitat.*

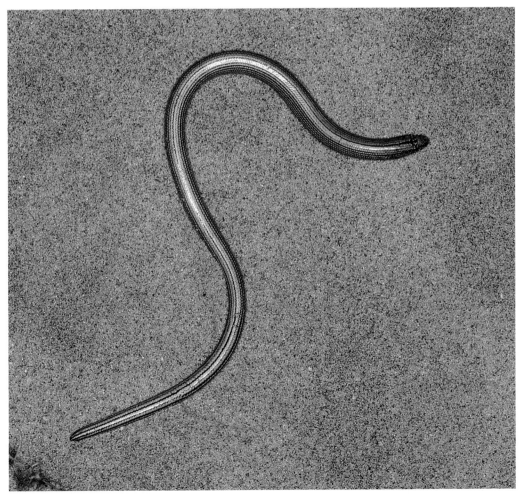

## FAMILY: CORDYLIDAE

The cordylid family is confined to the African continent. The 33 species are heavily armoured with thick bony plates, underlying equally tough scales. These species are well suited to life in dry, barren, predominantly rocky areas. Many species are popular and interesting subjects for the vivarium, but may prove difficult to breed.

SUBFAMILY: Cordylinae
(Zonures or girdle-tails)

### Giant zonure; sungazer
*Cordylus giganteus*

**Distribution:** Southern Africa.
**Length:** 35-38cm(14-15in).
**Sex differences:** Probing may be necessary to determine sex.
**Diet:** Locusts and 'pinkie' mice or may accept raw meat or liver. Vitamin and mineral supplements are essential.

Below: *The dull colours of the giant zonure,* Cordylus *giganteus, enable the lizard to blend into the barren landscape.*

**Ideal conditions:** Semi-desert or desert vivarium, measuring at least 90x45cm(36x18in) for a pair. Lay down a newspaper base for these untidy lizards. A 'hot spot' is essential, although they tolerate a wide temperature range of 15-35⁰C(59-95⁰F).
**Hibernation:** No.
**In captivity:** They adjust extremely well to captive conditions and live for many years. If you are fortunate enough to breed these viviparous lizards, they will produce up to three 10cm(4in) young. Unlike the sandy brown adults, the juveniles are brightly coloured, with yellow throats and snouts.

**Other species of interest:** The sungazer is the largest and most spiny cordyline; other species require a smaller vivarium, but similar conditions. Jones' zonure, *C. jonesi*, a spiny reddish lizard, is the smallest member of this subfamily at 12-16cm(4.7-6.2in). It frequently appears on dealers' lists and is easy to maintain. The most attractive species is the blue-spotted zonure, *Cordylus caeruleopunctatus*. Being a mountain dweller, it prefers slightly lower temperatures. The armadillo lizard, *C. cataphractus*, Warren's zonure, *C. warreni*, and the common girdle-tail, *C. cordylus*, to name but a few, occasionally become available and all are generally recommended.

### Red-tailed flat rock lizard
*Platysaurus guttatus*

**Distribution:** Southern Africa.
**Length:** 18-24cm(7-9.5in).
**Sex differences:** Males are highly colourful, with an iridescent blue and green head and body, reddish tail and silvery blue underside. In contrast, females are brown, with a dirty yellow vertebral stripe.
**Diet:** Insects and pinkie mice.
**Ideal conditions:** Semi-desert, with plenty of rocks and hiding places. Temperature range 18-32⁰C(64-90⁰F).
**Hibernation:** No.
**In captivity:** Although initially nervous and shy, these lizards soon settle down and breed regularly. Flat rock lizards deposit two to four eggs several times a year. The young, which hatch out after 40-50 days, resemble the female in coloration and feed on small crickets, caterpillars and flies.

Above: *Its striking*    Platysaurus guttatus
*coloration has made*    *a very popular pet.*

The ten species of *Platysaurus* flat rock lizard differ from zonures in that they are highly coloured, lack armour plating and are very flat, enabling them to climb into narrow crevices. All the flat rock lizards require similar conditions and care.

**Other species of interest:** The impressive imperial flat rock lizard, *P. imperator*, reaches over 35cm(14in) in total length and inhabits northeastern Zimbabwe and Mozambique. The Cape flat rock lizard, *P. capensis*, a smaller 20cm(8in)-long species, is found in very barren areas of southern South Africa.

SUBFAMILY: Gerrhosaurinae (Plated lizards)

**Yellow-throated plated lizard**
*Gerrhosaurus flavigularis*

**Distribution:** From Sudan to South Africa.
**Length:** The smallest species, measuring 35-42cm(14-16.5in).
**Sex differences:** Probing may be necessary to determine sex.
**Diet:** Insects and some vegetable matter.
**Ideal conditions:** Semi-desert, with many robust rocks, logs and hiding places. The preferred temperature range is around 21-35°C(70-95°F).
**Hibernation:** No.
**In captivity:** Initially shy, but settles down to make an excellent captive. Plated lizards are equipped with bony scales that provide

protection against the sun and predators. They are egg layers, but seem somewhat reluctant to mate and deposit their three to five large eggs.

**Other species of interest:** The largest species is the South African plated lizard, *G. validus*, which attains nearly 70cm(27in). If provided with a large sturdy vivarium, it settles down well in captivity. The most desirable species is the Sudanese plated lizard, *G. major*, from central and southern Africa, which is slightly smaller than *G. validus* at around 55cm(21.5in). These two species will eat mice, other small rodents and birds' eggs. A number of gerrhosaurids from South Africa called seps (*Tetradactylus*) have reduced limbs, giving them a snakelike appearance. They are worth trying to keep in captivity; they are unusual and interesting and only require a relatively small vivarium.

Below: *This head shot*    *shows the lizard's*
*of* Gerrhosaurus    *pointed snout, ideal for*
flavigularis *clearly*    *burrowing activities.*

## FAMILY: GEKKONIDAE

This vast family includes some of the strangest and most interesting reptiles. Until recently, many species had never been observed, let alone kept in captivity, but ever-increasing numbers of geckos from around the world are gradually becoming available. This is mainly due to their uncomplicated captive and breeding requirements and tolerance of a wide range of conditions. Indeed, most specimens offered by dealers today are captive-bred.

Most species are able to shed their tails at the slightest touch, so take great care when attempting to handle geckos. They are well known for their ability to climb up vertical surfaces or travel upside down along horizontal surfaces, using the modified pads underneath their toes. (Some species do, however, display the more conventional claw and toe.) Do be sure to provide them with an escape-proof vivarium.

Although many lizards can hiss or - to a certain extent - grunt, geckos are the only truly vocal lizards, emitting squeaks and yelps of varying tones and frequencies. The exact meaning of each sound uttered by a gecko is uncertain, but one cry is probably a mating call and another is territorial.

Geckos are vocal because they are predominantly nocturnal. Bright coloration or body displays during mating or when asserting dominance to secure a territory, for example, would be ineffective in the dark. As most geckos have excellent hearing, communicating with other geckos by sound

is ideal. To a certain extent, they can also warn of dangers close by, such as a predator.

Having evolved from nocturnal species, day geckos - for example *Gonatodes*, *Phelsuma* and *Sphaerodactylus* species - are also vocal. They have never lost their voices and use them to advantage within the thick forest undergrowth in which they live.

As the family is so extensive, it has been divided into four subfamilies, each one having a distinct appearance or particular behavioural characteristics.

SUBFAMILY: Diplodactylinae
Members of this subfamily originate from Australia, New Zealand and some Pacific Islands, and are found in savanna or tropical forest. Distinguishing features include their extraordinary shapes and sizes, and also the eyelids, which have fused to form a protective covering. At present, few species are encountered in captivity, which is a pity since they represent some of the most spectacular of the gekkonids.

### Leaf-tailed gecko
*Phyllurus platurus*

**Distribution:** Australia.
**Length:** 22-27cm(8.5-10.5in).
**Sex differences:** Males have well developed pre-anal pores.
**Diet:** Insects, small rodents and other lizards.
**Ideal conditions:** A tall, spacious planted vivarium, with high humidity and plenty of branches for the lizards' arboreal habits.

Left: Phyllurus platurus, *the leaf-tailed gecko, admirably camouflaged for effective protection.*

Right: *The vertically split pupil is a clue to the nocturnal habits of* R. auriculatus, *from Australia and Asia.*

Below: *The leopard gecko,* Eublepharus macularis, *reserves fat in its tail for future use when food is short.*

Provide a minimum temperature of 24⁰C(75⁰F), but 30⁰C(86⁰F) is better.
**Hibernation:** No.
**In captivity:** A shy, rarely seen species in the wild, due to its superb camouflaging. In captivity, it is equally shy, but lively and easy to maintain.

**Other species of interest:** Although the strange leaf-tailed gecko is the most widely available Diplodactyline, other species are equally impressive. One of the world's largest geckos, the Caledonian or giant gecko, *Rhacodactylus leachianus*, from eastern Australia and the South Pacific Islands can attain nearly 37cm(14.5in). This docile but shy species occurs in tropical forests, where it preys on mice, small birds, insects and even fruit, such as banana. Its unusually stumpy tail and broad body give it a bizarre appearance, and the underside of the tail has specialized scales to provide the additional adhesion needed to support its large size. In captivity, the giant gecko needs a spacious

vivarium and appears to enjoy - even demand - attention from its owner. It very rarely bites, even if molested.

Other smaller, Australasian rhacodactylids worth seeking out are *R. chahoua* and *R. auriculatus*. All diplodactylines require similar care. They lay either one or two large, parchmentlike eggs (3.5cm/1.4in long in *R. leachianus*), often depositing them in nests excavated by the female. The hatchlings are relatively large and can be treated in the same way as adults. Like most geckos, they are predominantly nocturnal or crepuscular (becoming active at dawn and dusk).

SUBFAMILY: Eublepharinae
Eublepharines are the most primitive geckos, retaining true eyelids but lacking clinging pads. Their distribution is scattered throughout arid regions between the equator and the Tropic of Cancer, and all but a few species are terrestrial. Many species are seen in captivity, as they are easy to breed.

**Leopard or panther gecko**
*Eublepharus macularis*

**Distribution:** Southwestern Asia to northwestern India.
**Length:** Up to 24cm(9.5in).
**Sex differences:** Males are more robust and aggressive, with well-developed pre-anal pores.
**Diet:** Large insects, pinkie mice or even canned dog meat.
**Ideal conditions:** A spacious desert-type

vivarium with plenty of hiding places. A daytime temperature of around 28⁰C(83⁰F) is sufficient.

**Hibernation:** No, but lowering the temperature to 21⁰C(70⁰F) and reducing the photoperiod for several months in the winter will stimulate reproduction.

**In captivity:** Very easy to maintain and a good choice for the beginner. Leopard geckos breed regularly throughout the spring and summer. Since males tend to be highly territorial and aggressive, it is wise to keep a single male to a group of two to six females. During the breeding season, females lay between three and seven clutches, consisting of one to three eggs. Unlike other geckos, eublepharines lay soft-shelled eggs. These need to be kept damp throughout incubation, so provide a damp area in the vivarium where the females can deposit their eggs. These are incubated at 24-32⁰C(75-90⁰F) and hatch out in 35-90 days, depending on the species. Leopard gecko hatchlings are large and extremely beautiful, striped in black, yellow and white. They devour crickets, spiders and small earthworms.

**Other species of interest:** Other eublepharines belonging to the genera *Coleonyx*, *Holodactylus* and *Hemitheconyx* are equally recommended in captivity. The banded gecko, *Coleonyx variegatus* and the Texas banded gecko, *C. brevis*, are small

Above: *The thin, fragile skin of* Coleonyx variegatus *is very delicate. Be sure to handle this lizard with great care.*

geckos from southern USA, measuring 7-13cm(2.75-5in) long. A small, dry vivarium kept at a temperature of 21-27⁰C(70-80⁰F) is ideal for them. Hatchlings of *Coleonyx* and *Holodactylus* are initially rather delicate, but grow and reach maturity within 8-10 months, given the correct diet.

SUBFAMILY: Gekkoninae
This most widespread and successful gecko subfamily is found on every continent except Antarctica. It contains some very interesting, beautiful and bizarre species, many of which

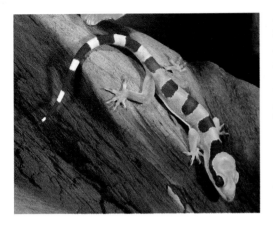

Above: *The markings of juvenile gekkonids are brighter than those of the adults. This is a young* Hemidactylus fasciatus *in captivity.*

are obtainable from reptile dealers. For descriptive purposes, this subfamily can be divided into two types: the typical gekkonines, such as the Tokay gecko, and the diurnal gekkonines, including the many *Phelsuma* species.

All gekkonine species lay between one and three calcareous eggs, so it is very important to add calcium to their diet. Before and during egg deposition, females will 'lick' powdered calcium and store it in endolymphatic glands located behind the jaw or synthesize it in the skin in the presence of ultraviolet light. Newly laid eggs are sticky

Left: *The large Tokay gecko,* Gekko gecko, *feels very much at home in the leafy surroundings of a humid vivarium.*

Below: Ptychozoon kuhli *is one of many geckos with special toe pads that enable them to climb the smoothest of surfaces with ease.*

and may adhere to the underside of leaves, rocks, logs or even to the vivarium sides. Attempts to separate them from any surface invariably result in damage, so either remove the whole leaf, rock or log, plus the egg for incubation or leave the egg in place and cover it with a ventilated plastic container. The baby gecko can then be removed on hatching. Another alternative is to provide short lengths of hollow bamboo tube for egg laying and transfer these to an incubation container kept at 24-32°C(75-90°F). Hatching begins 12-200 days later, depending on the species (see also *Breeding* page 53).

## Tokay gecko
*Gekko gecko*

**Distribution:** India, Southeast Asia.
**Length:** Up to 34cm(13.5in).
**Sex differences:** Males have pre-anal pores and a swollen tail base.
**Diet:** Large crickets, locusts, pinkie mice.
**Ideal conditions:** Either a 'hygienic' or a planted vivarium, with plenty of branches, rocks and hiding places. Temperature range 20-30°C(68-86°F).
**Hibernation:** No.
**In captivity:** Easy to maintain, but these ill-tempered geckos are capable of giving a painful bite. Wear gloves when handling large individuals. Not for the beginner.

**Other species of interest:** Other typical gekkonines tend to be less irascible, and many are smaller than the Tokay gecko. The list of available species is endless. Species from the genera *Hemidactylus*, *Cyrtodactylus* and *Tarentola* are a common sight in their native countries, by street lamps or catching insects in the houses. The Turkish, *H. turcicus*, common house, *H. frenatus*, Brooks, *H. brooki*, Kotschy's, *Cyrtodactylus kotschyi* and the Moorish gecko, *Tarentola mauritanica*, all make excellent, if somewhat timid, captives and breed regularly. All are less than 16cm(6.2in) long and, since they can be housed in a container as small as a plastic margarine tub, are ideal for hobbyists with limited space.

The most eye-catching gekkonines are Kuhl's gecko, *Ptychozoon kuhli*, and the fan-footed gecko, *Ptyodactylus hasselquistii*. In Kuhl's gecko, the dull mottled coloration,

Left: *This beautiful day gecko is at home among trees and shrubs. Here, it can search inquisitively for insects and take nectar from the flowers.*

webbed feet and flaps of skin running from the neck to the end of the tail enable it to become almost invisible on rocks or logs. It is also possible that such ornamentation may be used in gliding. For example, the flaps of skin between the legs can be stretched out when the legs are straightened, acting as a 'parachute' when the lizard attempts to reach another tree or if it accidentally falls. The fan-footed gecko has huge, triangular toe pads, enabling it to climb the smoothest surface.

## Day geckos
*Phelsuma* sp.

**Distribution:** Madagascar, other Indian Ocean Islands and the East African coast (introduced).
**Length:** From 8cm(3.2in) in *Phelsuma ornata* to 26cm(10in) in *P. madagascarensis*.
**Sex differences:** Males are highly territorial, have well-developed pre-anal pores and, in some species, are more colourful or smaller than females.
**Diet:** Insects; also fruit, nectar and calcium (either added to the food or place a small dish of ground cuttlefish bone in the vivarium).
**Ideal conditions:** The majority of gekkonine species are arboreal and require a tall planted vivarium, frequently misted to maintain high humidity, and a basking area. Some of the smaller species can be kept in small plastic containers. Ultraviolet light is essential for calcium synthesis. Temperature range 19-31⁰C(66-88⁰F).

**Hibernation:** No.
**In captivity:** In a suitable vivarium, these geckos are easy to maintain and breed.

Day geckos, brilliantly coloured in iridescent reds, greens, blues and gold are aptly described as the 'jewels' of the herptile world. They differ from other geckos in having rounded pupils and being totally diurnal, which makes them highly interesting captives. All the many available species are captive-bred, as wild populations are strictly protected. The initial outlay for these geckos can be quite substantial, but as they are easy to breed and much sought-after by collectors, they are well worth the cost.

**Other species of interest:** All the smaller species can be recommended for beginners and experienced hobbyists alike. The most commonly available are the stunning Mauritius, *Phelsuma cepediana*, gold-dust, *P. laticauda*, lined, *P. lineata*, dull day gecko, *P. dubai*, and Mauritius ornate, *P. ornata*, the smallest of the genus. In fact, many of the 29 known species are readily available, plus a great many subspecies. The Madagascar giant, *P. madagascarensis*, requires a more spacious cage than other species, but is equally spectacular and easy to maintain.

SUBFAMILY: Sphaerodactylinae
This subfamily contains the most specialized, advanced and smallest lizards in the world. Although many are short-lived and few are commonly supplied by dealers, they are in demand because they breed regularly and are ideal for collections with limited space.

**Caribbean gecko**
*Sphaerodactylus elegans*

**Distribution:** The Antilles.
**Length:** Rarely larger than 4cm(1.6in).
**Sex differences:** Male coloration is brighter.
**Diet:** Small crickets, flies and caterpillars.
**Ideal conditions:** A few specimens can be kept together in an ice-cream container, but a small colony would look attractive in a tall, planted vivarium. These geckos prefer high humidity and temperatures in the range of 24-30°C(75-86°F).
**Hibernation:** No.
**In captivity:** Very easy, but keep a close eye on potential escape points in the vivarium.

**Other species of interest:** *Sphaerodactylus cinereus* and *S. parthenopian* from the West Indies are very similar in appearance and behaviour to *S. elegans*. The genus *Gonatodes* is slightly larger, attaining 6-8cm(2.4-3.2in). The banded gonatode, *G. vittatus*, and the Trinidad gonatode, *G. humeralis*, are the most commonly encountered species.

Above: Phelsuma *species, (this is* P. quadriocellatus*) have round pupils and stunning coloration. They are all diurnal.*

Below: *Although barely 8cm long, the agile* Gonatode humeralis *is still a larger member of the Sphaerodactylinae.*

All species are diurnal or crepuscular and breed throughout the year, the female laying one large calcareous egg four to six times each year. The egg hatches out after 50-75 days at a temperature of 24.5°C(76°F) and the resulting 1.5-2.5cm(0.6-1in) hatchling is very delicate, but will attain maturity within six to eight months. Provide a diet of fruit flies and tiny crickets.

## FAMILY: IGUANIDAE

Iguanas are some of the most popular vivarium occupants and a good selection is usually available. Some of the smaller species are ideal for the beginner, but larger types, such as the rhinoceros iguana, *Cyclura cornuta*, can inflict serious injury with their powerful limbs and are best left to zoological collections. This very large family is considered under three subfamily headings.

SUBFAMILY: Iguaninae

### Green or common iguana
*Iguana iguana*

**Distribution:** Central America and northern South America.
**Length:** Can attain 200cm(78in) but usually around 130cm(51in).
**Sex differences:** Males have larger heads and brightly coloured limbs.
**Diet:** Herbivorous. May accept insects when young.
**Ideal conditions:** A huge enclosure, with plenty of branches. A heated greenhouse would be better. Temperatures no less than 27°C(80°F) and up to 32°C(90°F). Juveniles will refuse food at lower temperatures.

Above: *The prehistoric* Iguana iguana *has strong claws and green camouflage, ideal for a life in the trees.*

**Hibernation:** No.
**In captivity:** Settles down well in captivity and will breed, if given extensive space and the right conditions.

**Other species of interest:** The green iguana is one of the most spectacular lizards available. Depending on the country of origin, it varies in colour, from shades of grey to striking emerald green, with black bands along the tail. Smaller iguanines, such as the insectivorous desert iguana, *Diposaurus dorsalis*, and the herbivorous chuckwalla, *Sauromalus obesus*, grow to about 25-45cm(10-18in) and thrive in a desert vivarium with high temperatures. Breeding is rare, except in spacious enclosures, where a single male will mate with several females. About 25-60 eggs are deposited in damp sand and these hatch in 55-65 days at a temperature of 27-31°C(81-88°F). The juveniles of all species are fairly large and can reach maturity in two years.

Equally popular are the curly-tailed iguana, *Liocephalus barahonensis*, and the ornate curly-tailed iguana, *L. personatus*, from

Haiti and other Antilles. They are easy to maintain as colonies in a relatively small planted vivarium equipped with a basking area. Both species grow to 15-17cm(6-6.5cm) and may breed several times a year, the female laying up to 20 small eggs at a time.

SUBFAMILY: Sceloporinae

**Western fence lizard**
*Sceloporus occidentalis*

**Distribution:** Western USA.
**Length:** 15cm(6in).
**Sex differences:** Males have blue areas on the belly and throat.
**Diet:** Insects and spiders.
**Ideal conditions:** A dry vivarium, with plenty of rocks, logs and branches. It enjoys a 'hot spot' and daytime temperatures around 30°C(86°F).
**Hibernation:** No.
**In captivity:** These small lizards are ideal for the beginner.

**Other species of interest:** Many other fence and spiny lizards of the genus *Sceloporus* are available from time to time. The granite spiny, *S. orcutti*, and the sagebrush lizard, *S. graciosus*, are perhaps the most attractive. All species from the USA require similar care. Some species are viviparous, bearing three to five young after a long gestation period, whereas others lay 10-15 eggs, which hatch out after 45 days at a temperature of 27°C(81°F). The young are relatively large.

The most unusual sceloporines are 14 species of desert-dwelling horned lizards, *Phrynosoma* sp., from the USA. They require a spacious vivarium and similar conditions to spiny lizards but, regrettably, they tend to be short-lived in captivity. Their diet consists of ants, crickets and mealworms, which they demand in huge quantities. They are either viviparous or oviparous, depending on the species, but are difficult to breed and therefore not recommended for beginners.

Two other sceloporines worth mentioning are the collared lizard, *Crotaphytus collaris*, and the beautiful leopard lizard, *C. wislizenii*, from the southern USA. Males are particularly brightly coloured. Both are strong and robust, well suited to a large and very hot, savanna-type vivarium, where they will feed greedily on insects, flowers, small mammals or even raw meat. Breeding is possible under ideal conditions, with 4-10 eggs being laid in spring. These take about 55 days to hatch at a temperature of 30°C(86°F). The hatchlings are very fond of spiders.

Above: *Spiny lizards are so-called because their keeled scales end in a point. This is* Sceloporus occidentalis *from the western USA.*

Left: Crotaphytus wislizenii, *the desert-dwelling leopard lizard, can grow up to 40cm(16in) long.*

SUBFAMILY: Anolinae

**Knight anole**
*Anolis equestris*
**Distribution:** Cuba and southeastern USA (introduced).
**Length:** 30-40cm(12-16in).
**Sex differences:** Males have two large scales behind the cloaca.
**Diet:** Insects and pinkie mice.
**Ideal conditions:** A tall planted vivarium, with a basking area and high levels of humidity. Provide a temperature range of 20-30°C(68-86°F).
**Hibernation:** No.
**In captivity:** Easy to maintain, but rather docile. May need feeding individually.

**Other species of interest:** Anoles are popular pets, especially in the USA, and several species are available. The knight anole is the largest and most docile, but all species require similar care. The green anole, or American chameleon, *A. carolinensis*, is an inexpensive, interesting lizard and rather more agile. Males are distinctly larger, display pink throat fans or dewlaps and may fight furiously over a female.

Breeding takes place throughout the year in all anoles, with the female laying one to four eggs beneath rocks or logs. The eggs hatch within 40 days and the resulting hatchlings are rather delicate, but grow quickly on a diet of small crickets and flies.

SUBFAMILY: Lacertidae

Lacertids are the typical lizards often seen in Mediterranean holiday resorts, basking on walls or near roads. The family is widespread over Europe, Africa and Asia and within this area the diversity in colour is immense. Such is the variation in individual species over different parts of the range that new subspecies have been formed, with as many as 36 in some wall lizards, *Podarcis* sp. In captivity, they are undemanding and easy to breed, but in recent years the number of species offered by dealers has declined dramatically, due to loss of habitat in the wild, governmental protection in the respective countries of origin and even overcollecting. Lacertids can be divided into two categories: small lizards, such as the *Podarcis* species, and green lizards.

**Italian wall or ruin lizard**
*Podarcis sicula*

**Distribution:** Italy and isolated regions in southern Europe.
**Length:** 18-25cm(7-10in).

**Sex differences:** Females are more striped and less robust, with a smaller head.
**Diet:** Insects and some fruit or vegetable matter.
**Ideal conditions:** A dry, warm vivarium, complete with basking area and ultraviolet light. Temperature range 21-30°C(70-86°F).
**Hibernation:** No, but lower the vivarium temperature by 5-7°C(8-13°F) during the winter.
**In captivity:** Adapts well to captivity and frequently breeds.

**Other species of interest:** Many other small lacertids regularly become available and all require good basking areas, as they are essentially sun-loving. The common wall lizard, *Podarcis muralis*, is a good beginner's

lizard, at home both in an indoor or outdoor set-up. There are over 30 small lacertid species in Europe alone, some of which are particularly colourful - the milos wall, *P. milensis*, Lilford's wall, *P. lilfordi*, and the sharp-snouted rock lizard, *L. oxycephala*, to name but a few. Surprisingly, even some of the more southerly species, such as the snake-eyed lizard, *Ophisops elegans*, and small species such as *Algyroides nigropunctatus* and *A. moreoticus*, will adapt to outdoor or greenhouse conditions, as long as they have adequate retreats and are well protected from severe winter temperatures.

Most species live in small colonies with one dominant male. Breeding can occur several times from spring until autumn. The female lays small clutches of two to eight eggs, which hatch out after 40-50 days at a temperature of 24°C(75°F). The hatchlings grow quickly if fed on a diet of fruit flies and small crickets.

The common lizard, *Lacerta vivipara*, is the only livebearing lacertid and, being extremely hardy, prefers an outdoor position with plenty of basking points and shade. It produces 3-12 miniature replicas of itself in the middle of summer.

The strangest species is the attractive six-lined or long-tailed lizard, *Takydromus sexlineatus*, from Thailand which, as its

Above: *Lacertids are a familiar sight in the Mediterranean and there are many subspecies. This is* P. sicula campestris.

Left: *Attractive variation can occur in anoles, as shown in this photograph of a blue Knight anole,* Anolis equestris.

Right: *Island populations of wall lizards, such as Vedra's* Podarcis pityusensis vedrae, *are often intricately marked.*

Left: *The sun-loving green lizard,* Lacerta viridis, *often climbs trees in search of a more favourable basking site or for a meal of ripened berries.*

Below: Lacerta lepida *is a large species, sometimes known as the jewelled lizard, because of the distinctive ornamental blue spots present on the animal's flanks.*

common name suggests, has a long tail as much as four times the length of its body. Until recently, this species was included in the subfamily Lygosominae but research has shown that it belongs with the Lacertidae.

**European green lizard**
*Lacerta viridis*

**Distribution:** Central and southern Europe.
**Length:** 30-40cm(12-16in).
**Sex differences:** Males are more brightly coloured and have a thicker tail base.
**Diet:** Invertebrates, pinkie mice and some fruit.
**Ideal conditions:** A dry vivarium with plenty of rocks and logs. Temperature range 19-27⁰C(66-81⁰F).
**Hibernation:** No, but lowering the temperature to around 12⁰C(55⁰F) for a few months in the winter will invigorate breeding behaviour.
**In captivity:** Easy to keep in a well-lit vivarium or even a greenhouse, but prone to skin disorders if not given access to sufficient natural light.

**Other species of interest:** Other green lizards may become available, usually from a captive-bred source. The Balkan green, *L. trilineata*, and the Iberian green, *L. schreiberi*, are similar in appearance and habits and respond to the same care as *L. viridis*, but

both require slightly higher minimum temperatures. The largest and most outstanding species is the eyed lizard, *L. lepida*, from Spain, Portugal, southern France and northwestern Africa, which may grow to over 80cm(32in). Although requiring similar conditions to other green lizards, it obviously needs a larger vivarium - especially for breeding - and it also takes larger foods, such as small rodents and birds. It is often long-lived in captivity, frequently reaching about 18 years old.

Green lizards regularly breed, with males fighting for a female. Large clutches of 10-25 eggs are laid - sometimes twice a year - and these can be incubated in a warm, humid container at about 30⁰C(86⁰F). Hatching begins 40-60 days later. The young lizards are relatively large and very colourful and reach maturity in 12-18 months.

## FAMILY: SCINCIDAE

Skinks represent the largest lizard family, with over 700 species known today. Although not as diverse in shape and size as the iguanids, they are among the easiest families to maintain and breed in captivity and are highly resistant to disease. There are both egg laying and livebearing species within the family; some egg layers are among the few lizards to protect their eggs.

As the family is so large, it is described under three subfamily headings.

SUBFAMILY: Lygosominae

### The gold skink
*Mabuya multifasciata*

**Distribution:** Burma, Malaya and Thailand.
**Length:** 18-25cm(7-10in).
**Sex differences:** Males are more robust in the neck and head region.
**Diet:** Insects, small snails, pinkie mice or even overripe fruit.
**Ideal conditions:** A warm vivarium with a substrate of loose material on the base and a small basking area. Daytime temperature around 27°C(81°F), but cooler at night.
**Hibernation:** No.

Below: *A head shot of the skink* Mabuya multifasciata *shows the external ear opening, a feature absent in snakes.*

**In captivity:** An easy and undemanding species that will live for many years.

**Other species of interest:** The subfamily Lygosominae takes its name from the Greek ('Lygo': a long flexible twig and 'soma': body). Lygosomids are very diverse in shape, some species having a typical lizard appearance with well-developed limbs, some having much reduced limbs, while others have a snakelike appearance. The subfamily is widespread, consisting mainly of African and Asian skinks, plus a few from southern Europe and South America. The only European representative is the minute burrowing snake-eyed skink, *Ablepharus kitaibelii* which, unlike many lygosomids, produces eggs and prefers moist habitats. The gold skink, the African five-lined skink, *M. quinquetaeniata* and the large orange-throated skink, *M. macularis* are commonly offered by dealers, usually as wild-caught specimens. Lygosomids are very easy to breed and, as most species are diurnal, make interesting vivarium subjects.

SUBFAMILY: Scincinae

### Blue-tailed skink
*Eumeces skiltonianus*

**Distribution:** Western USA
**Length:** Up to 16cm(6.2in), usually smaller.

**Sex differences:** Difficult to determine, although the male may have a broader base to the tail.
**Diet:** Mainly insectivorous.
**Ideal conditions:** Prefers a moist, woodland vivarium, with plenty of hiding places and a basking area. Temperature range 19-27°C(66-81°F).
**Hibernation:** Yes.
**In captivity:** A very attractive and easy species, suitable for an indoor, outdoor or greenhouse vivarium.

**Other species of interest:** This is the largest subfamily found throughout the Americas, Europe, Africa and Asia. American species are egg layers and the females actually guard the eggs. Many species have much reduced limbs or are snakelike in appearance. In *E. skiltonianus*, the striking blue tail fades with age, although the vivid dorsal patterning remains. There are many other species of *Eumeces* and some of the North American species can be described as hardy, but it is important to check exactly where your lizard originated from. *Eumeces* lays up to 18 eggs, which can either be incubated in a separate container at 30°C(86°F) or left for the female to care for.

Other genera of Scincinae found in reptile shops include the burrowing *Scincus* species and the snakelike *Ophiomorus* species, both from Western Asia, and the Mediterranean three-toed skink, *Chalcides chalcides*. All make good vivarium subjects and are generally livebearing, giving birth to 3-23 young. Provide relatively dry conditions with plenty of loose material such as chipped bark to crawl through.

SUBFAMILY: Tiliquinae

**Blue-tongued skink**
*Tiliqua gigas*

**Distribution:** New Guinea and islands of the Banda Sea.
**Length:** 50cm(20in).
**Sex differences:** Males have a larger head and thicker tail base.
**Diet:** Insects, snails, small rodents, fruit.
**Ideal conditions:** A simple but spacious set-up with a few rocks and a hide. Temperatures in the range 19-30°C(66-86°F).
**Hibernation:** No.
**In captivity:** An easy species that breeds readily and is long-lived.

Below: *The blue tail of the juvenile five-lined skink,* Eumeces fasciatus, *serves to distract a predator. If the lizard is caught, it can discard its tail and make its escape.*

As the common name suggests, the distinguishing feature of these skinks is their vivid cobalt blue tongue. The precise reason for this blue coloration is unknown, but a display in courtship or of aggression are the most likely explanations.

**Other species of interest:** The subfamily is confined to Australia, New Guinea and some Pacific Islands and contains some of the most impressive - and thus extremely popular - of all the herptiles. Other tiliquines regularly seen in captivity include the Australian blue-tongue, *T. scincoides*, the unusual Gerrard's blue-tongue, *T. gerrardii*, and the strange stump-tailed skink, *Trachysaurus rugosa*. The fat stumpy tail of this latter species gives it

the appearance of having two heads - an effective defence mechanism.

The arboreal Solomon Islands giant skink, *Corucia zebrata*, grows over 60cm(24in) long and gives birth to a single 15cm(6in) skink. It is the largest skink in the world and is now becoming available to hobbyists.

Tiliquines tend to be rather aggressive, especially the males, and are best housed in individual vivariums. It has now been proved that tiliquines breed in late spring to early summer in the Northern Hemisphere, irrespective of vivarium conditions. All species are livebearers, giving birth to as many as 25, 13cm(5in)-long young. Remove these immediately to prevent cannibalism and give them the same care as the adults.

Above: *The waxy skin of the giant blue-tongued skink,* Tiliqua gigas, *reduces water loss in arid conditions.*

Right: Trachysaurus rugosa, *the stump-tailed skink, clearly seen in a warning display, typical of these blue-tongued skinks.*

## FAMILY: TEIIDAE

This relatively large family is the American equivalent of the Old World lacertids. They inhabit many different types of terrain, some species having very reduced limbs and leading a subterranean existence. They vary considerably in size, ranging from 7.5-120cm (3-48in), and in scale formation; some species look quite skinklike.

### Six-lined racerunner
*Cnemidophorous sexlineatus*

**Distribution:** Southeastern USA.
**Length:** 18-22cm(7-8.5in).
**Sex differences:** Males have a blue throat and belly.
**Diet:** Insects.
**Ideal conditions:** A warm, dry vivarium with a loose substrate and plenty of hiding places. Temperature around 27°C(81°F).
**Hibernation:** No, but cool down by 5-7°C (8-13°F) during the winter.
**In captivity:** Interesting lizards that do well in captive conditions, given sufficient warmth and plenty of cover.

**Other species of interest:** Many species of racerunners, or whiptails, occur in the USA and Central American countries and all are similar in size and shape, although the dorsal patterning may vary from spots to stripes.

Above: *In the wild, the very agile* Cnemidophorus sexlineatus *takes full advantage of its streamlined shape.*

The six-lined racerunner is one of the most agile lizards, attaining speeds of up to 28kph (17.5mph), so you must be quick when attempting to catch them. Some whiptails, such as *C. uniparens, C. exsanguis* and *C. tessalatus*, are parthenogenetic (they lay unfertilized eggs that produce only females) and males are unknown or extremely rare.

Other teiids sometimes available include the jungle runners, *Ameiva* sp., robust, colourful lizards that require a spacious vivarium with plenty of branches, sturdy plants and warm, (30°C/86°F) humid conditions. Members of the genus *Euspondylus*, on the other hand, are often called snake teiids because of their tiny limbs and shiny scales. They are easy to maintain in a small vivarium.

All teiids are oviparous, producing a single clutch of up to eight eggs in racerunners and ameivas. Remove the eggs immediately so that they are not devoured by the adult and keep them in a humid atmosphere at about 27°C(81°F); they will start to hatch 55-70 days later. Young racerunners are particularly well marked, some species having bright blue tails. Small insects form their basic diet and they can attain maturity in 12 months.

Above: *The lidless catlike eyes of* Xantusia vigilis *are typical of a species chiefly active at night.*

Left: *Night lizards, such as* Xantusia henshawi, *are very attractive, but not widely available.*

## FAMILY: XANTUSIDAE

This family of four genera is confined to the USA, Mexico and Central America. Their common name - night lizard - refers to their usually nocturnal behaviour, although some individuals are diurnal. Some species are geckolike in appearance. Although attractive, only a few species are occasionally available.

### Desert night lizard
*Xantusia vigilis*

**Distribution:** Southwestern USA.
**Length:** 8-10cm(3.2-4in).
**Sex differences:** Males have well-developed femoral pores.
**Diet:** Insects and spiders.
**Ideal conditions:** A small vivarium with an abundance of rocks and logs. Temperature range 19-27°C(66-81°F).

**Hibernation:** No, but reduce temperature by 5°C(8°F) during winter.
**In captivity:** This species can be active during the day, but is very secretive. Otherwise easy and undemanding.

**Other species of interest:** The only other species sometimes seen in captivity are the granite night, *X. henshawi*, and the Cuban night lizard, *Criosaura typica*. Both are small lizards, leading solitary lives and showing no aggression towards other lizards or to their keeper. Take care when handling them as the tail is very fragile.

All species are viviparous, giving birth to no more than three young after a 90-120 day gestation period. The newborn lizards are delicate at first, but on a diet of fruit flies or greenflies, liberally dusted with multivitamin powder, they quickly reach maturity.

# Part two: Snakes

Just mention the word 'snake' and many people will recoil in horror. It stirs up memories of old wives' tales and conjures visions of giant, hooded snakes with poison-laced fangs, a stinging tail or 'Medusa' eyes. In fact, snakes are misunderstood creatures. It is true that there are dangerous species, but without snakes, pests and disease carriers, including mice and rats, would cause havoc in poor agricultural and tropical areas. People certainly die from snake bites and even small, non-poisonous snake species can be aggressive but, with very few exceptions, this behaviour is displayed only when the snake is molested or otherwise antagonized.

Wild snakes are, in fact, very shy and secretive and frequently the target of persecution. In the southern USA, for example, the infamous rattlesnake hunt and chopping competitions have been internationally condemned. Millions of snakes are killed annually for food - in Japan and some Southeast Asian countries they are considered a delicacy - and many others die accidentally on the roads.

Approximately 2,700 species of snake are found throughout the world, apart from in polar regions, on some islands - for example, Ireland and New Zealand - and on isolated islets separated from land masses before snakes evolved. Unlike lizards, snakes have not been able to adapt fully to man's advancing concrete world, although they appear to be 'creeping' slowly into fringe areas of this synthetic jungle, such as gardens, parks or waste grounds.

Globally, snake populations are threatened as a result of habitat loss, persecution and overcollecting. Particularly vulnerable are those species from remote islands, where the introduction of 'supertramp' animals, such as rats, cats, pigs and goats - along with humans - has decimated entire populations. For example, *Casarea dussumieri* from Round Island in the western Indian Ocean was reduced to the dangerously low population of 54, but this is gradually rising thanks to captive breeding projects and now stands at around 100 individuals.

Ironically, as we learn more about the unique characteristics of snakes, they are becoming some of the most popular alternative pets. Increased research has led to improvements in vivarium technology and today, even species once thought to be 'difficult' can be easily maintained and bred in captivity. Many species in modern collections are captive-bred, which has reduced the pressure on wild populations, and these specimens are more adaptable and less aggressive than their wild counterparts.

# WHAT ARE SNAKES?

Snakes belong to the suborder Serpentes (sometimes called Ophidia), which is one half of the most successful reptile order Squamata. Lizards belong to the other half, the Sauria. Both snakes and lizards evolved from a prehistoric branch called Eosuchia nearly 300 million years ago, but present-day snakes did not begin to separate and evolve their streamlined shape until just 130 million years ago. Some of today's snakes, such as the colubrids, are the result of less than 20 million years of evolution.

When snakes broke away from lizards, gradual changes took place, the most obvious being the loss of the limbs, eyelids and eardrum. Even today, there is evidence of the snake/lizard connection; remnants of hind limbs in the form of vestigial claws near the tail base are present in some primitive snakes, such as boas. Similarly, many advanced lizards are in the process of losing - or have already lost - their limbs. In terms of shape, snakes are not diverse, but they show a vast array of colours and sizes.

## The skeleton

The skeleton of a snake is a wonderful example of adaptation. Its principal function is to facilitate movement but, depending on the species, it can be adjusted to feeding, defence or courtship behaviour. The main trunk of the snake consists of up to 500 vertebrae, each containing a pair of ribs. This arrangement allows snakes to explore and take advantage of many habitats, however hostile. The sternum and the shoulder girdle are absent (but the latter is always present in lizards), and the pelvic girdle is found only in primitive snake families. In many species, the ribs can be flattened out to create a wider area for trapping the sun's energy or for escaping under crevices and rocks. Flattening is also used as a defensive mechanism, because it makes certain snakes appear 'larger than life'. Cobras, (*Naja*) and hognosed snakes (*Heterodon*) are just two that take advantage of this ability.

The skull is joined to the first spinal vertebra and its thickness and shape depend upon the snake's habits. For example, the burrowing worm snakes, *Typhlops*, have developed a thick blunt skull, whereas arboreal snakes have a light, flexible one.

## Movement

Movement in snakes is variable. The most typical form of locomotion is 'serpentining' and involves the snake in continual movement from side to side. At the same time, it gains momentum by pushing on any irregularity on the ground. Some smaller snakes display 'concertina' movements, compressing the curved body from the tail end like an accordian and then pushing it out from the neck. Larger or heavily built snakes have sizable underside scales that continually ripple as a result of muscular action. These push against any rough surface, producing a caterpillar-like motion. Another form of locomotion unique to certain desert vipers, is called sidewinding, in which the snake travels in a series of lateral movements with only two parts of the body in contact with the ground at any one time. It appears to be pointing in one direction, but in fact progresses at an angle. Arboreal species are a

## Snake anatomy

*The interaction between the tongue and the Jacobson's organ is one of the most well developed and sensitive in the reptile world.*

*Most snakes can see undefined shapes, but some arboreal species have acute vision. The eyes are covered by a transparent scale.*

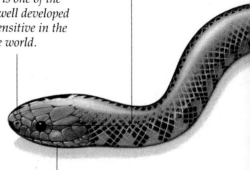

*The jaws can be voluntarily dislocated, enabling the snake to swallow massive prey items.*

good example of snakes that are able to use all types of movement (except sidewinding) or even combine them.

## The teeth

The teeth vary according to the family, but in most cases they are directly connected to the outer edge of the jaw bone. They tend to be small, pointed and curved back, enabling the snake to retain a better grip on its prey. The most specialized dentition is found in venomous snakes. Here, modified saliva,

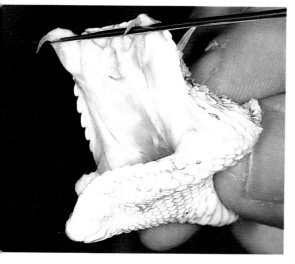

Left: *Like other members of the family, the horned viper,* Cerastes cerastes, *has hollow, hinged teeth with a 'syringe' mechanism that forces venom into the bloodstream of its prey.*

Above: *The snake's forked tongue flicks in and out all the time, gathering information from its surroundings. The mouth need not be open, as the tongue extends through a space below the snout.*

*The snake's long, limbless, streamlined body is ideal for seeking out holes, climbing trees or 'swimming' in the sand.*

*Snakes continue growing throughout their lives and periodically slough off their old skin.*

## Snake families and their distribution

| Family | Number of species | Common name and characteristics |
|---|---|---|
| Acrochordidae | 2 | Wart snakes. Aquatic snakes confined to Southeast Asia and some remote islands of that region. |
| Aniliidae | 9 | Pipe snakes. Brilliantly coloured snakes from southern Asia and northern South America. |
| Boidae | 110 | Boas; Pythons. This well known family consists of four subfamilies and contains the largest snakes in the world. Found in warmer regions worldwide. |
| Colubridae | 1230 | Typical snakes. The largest family is divided into 10 subfamilies and contains some of the most widespread and commonest snakes. |
| Crotalidae | 130 | Rattlesnakes; Pit vipers. Large, poisonous snakes found throughout the Americas and southern Asia. |
| Elapidae | 160 | Cobras (including the mamba, krait and coral snake). Contains some of the world's most dangerous and beautiful serpents. Distributed throughout the Americas, Africa, Asia and Oceania. |
| Hydrophiidae | 55 | Sea snakes. Almost entirely marine snakes with powerful venom. Found in warmer sea and coastal waters of the Pacific and Indian Oceans. |
| Leptotyphlopidae | 50 | Thread snakes. Minute wormlike burrowers from Southwest Asia, Africa and the Americas. |
| Typhlopidae | 400 | Blind snakes. Small, primitive reptiles with poorly developed eyes. Found throughout subtropical and tropical regions, as well as South Africa, southern Australia, southern Europe and western Asia. |
| Uropeltidae | 48 | Shield-tailed snakes. Small and related to the pipe snake. Confined to the Indian continent, including Sri Lanka. |
| Viperidae | 90 | True vipers. Mainly livebearing, venomous snakes, that occur as far north as the Arctic Circle and commonly found in Europe, Asia and Africa. |
| Xenopeltidae | 1 | Sunbeam snake. A unique and uncommon serpent, with highly polished scales. Widely distributed throughout Southeast Asia. |

Above: *With their flexible bodies, and unhampered by any limbs, snakes are masters of the trees. Here, an egg-eating snake (Dasypeltis) raids a warbler's nest.*

the snake to extend its jaw over prey as much as three or four times bigger than its own head. The prey is gradually pulled down the gullet by strong neck muscles as the highly elastic skin expands. Other muscles force open the windpipe to allow breathing. In fact, the entire internal structure of a snake is geared to feeding: the lung(s) are long and tubelike, the heart, liver and kidneys are flattened and the ribs are movable.

### The skin

Although all snakes have dry waterproof skin, the texture and scale size can vary from one species to another, depending on the snake's habitat. Minimal friction is an advantage to arboreal, subterranean (burrowing) or aquatic species and they are more likely to have smooth scales. Snakes from arid regions have thicker epidermal layers and rougher scales to give extra leverage on relatively smooth ground, such as sandy soil.

### The eyes and ears

Instead of eyelids, a clear protective scale covers the eyes of a snake. Good eyesight would be useless to burrowing snakes, so they have evolved small eyes with extremely poor vision. On the other hand, tree snakes of the genus *Dryophis* have acute vision that enables them to detect the slightest movement. Some snakes can distinguish colours, but most terrestrial snakes have relatively good black-and-white vision. They can see shades from white to light grey, dark grey and black, with varying degrees of clarity. It has been proved that a rattlesnake can see a moving object over 400m(1,290ft) away. However, even with good vision, snakes cannot distinguish stationary objects by eyesight alone. Like the lizards, they have developed a highly specialized sense - the Jacobson's organ. As the forked tongue is flicked in and out, it gathers microscopic odorous particles and deposits them into two highly sensitive, membrane-lined cavities in the roof of the mouth (Jacobson's organ). These particles are analyzed and the information is passed to the brain, enabling the snake to ascertain whether mate, enemy or food are in the vicinity. Some snakes have even more efficient sensors. Certain boids and pit vipers, such as rattlesnakes, have pits

known as toxin or venom, is secreted from glands and travels down grooves or through hollow tubes in the teeth. There are three main groups of venomous snakes:

**1** The back-fanged snakes (certain colubrids) have grooved teeth, located towards the back of the jaw. For their venom to be effective, these snakes must get a good grip and chew.

**2** The rigid front-fanged snakes (Elapidae) bite and instantaneously secrete a neurotoxin venom via a duct through the hollow elongated fangs. This penetrates the wound and attacks the victim's nervous system.

**3** The hinged front-fanged snakes (Viperidae and Crotalidae) have exceptionally long fangs (up to 5cm/2in in the gaboon viper, *Bitis gabonica*). When not in use, these lie flat, pointed backwards, ready to spring into action when the muscle of the jawbone, to which they are connected, rotates. The haemotoxin venom is forced at high pressure down hollow ducts and 'injected' deep into the victim's bloodstream. Here, it destroys red blood cells and causes haemorrhages.

### The jaws

When it comes to swallowing whole items of prey, even relatively large ones, the snake has few competitors. The jaws are independently mobile and loosely joined to one another - and to the skull - by elastic ligaments. In addition, the lower jawbone is split in half and also joined by ligaments. This enables

## Jacobson's organ

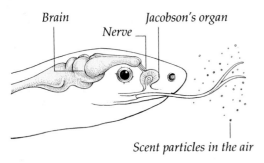

*Scent particles in the air*

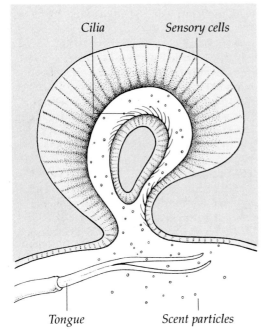

*Tongue*          *Scent particles*

Above: *A snake is often seen to flick its tongue repeatedly in and out. It is a myth that the tongue is laced with poison; in fact, its function is very complex. It may point into the air or just brush along the ground. By doing this, the snake is testing its surroundings. The tongue collects microscopic particles and, as they recoil, the forks brush - or even* enter - *the hollow cavity in the roof of the mouth, known as the Jacobson's organ. Situated within this organ are millions of hairlike projections onto which the scented particles are deposited. The particles are then analyzed, the information is passed to the brain and thus the snake is able to decide whether the area is safe and whether there is prey nearby.*

located below the eyes that are able to detect extremely slight changes in temperature - the body heat of a mouse, for instance. Thus, the snake is able to determine its exact distance from the heat source.

Another distinguishing feature is the lack of an external tympanum, or ear opening. Contrary to popular belief, however, snakes are not totally deaf and can respond both to ground vibrations and low-frequency airborne vibrations.

### Record breaking snakes

People are always fascinated by the sizes that snakes can attain - even if these statistics are sometimes vastly exaggerated! The two giants are the anaconda, *Eunectes murinus*, and the reticulated python, *Python reticulatus*. The former is by far the bulkiest of snakes, some individuals, weighing as much as 226kg(500lb). The reticulated python has been cited as the longest snake, with reliable reports of individuals measuring 10.2m(32.9ft). This snake grows to 6m(20ft) or more with greater regularity than any other snake. In stark contrast, the diminutive West Indian thread snake, *Leptotyphlops bilineata* rarely reaches 11.5cm(4.5in).

Snakes are also renowned for the proportionally massive foods they are able to swallow. The powerful members of the Boidae family - boas and pythons - are capable of swallowing chickens, pigs, smaller crocodiles, antelope and, very occasionally, humans. The king of the snakes in this respect is undoubtedly the African rock python, *P. sebae*: in 1955 an Impala antelope weighing over 58.8kg(130lb) was removed from a 5m(16ft) snake, the antelope being nearly as heavy as the snake itself.

Longevity is another feature of the larger, less active snakes that have fewer enemies. The common boa, *Boa constrictor*, from the Americas has been known to reach the ripe old age of 40 years.

Speed plays an important part in the survival of the smaller snakes. Otherwise, being so conspicuous in their natural habitat, they would easily fall prey to their enemies. The fastest snake is the deadly black mamba, *Dendroaspis polylepis* which, if threatened or angry, can attain short bursts of speed of 12-16kph(8-10mph). The grass snake, *Natrix natrix*, can travel at about 6.4kph(4mph).

## Regulating body temperature

Snakes are ectotherms, which means that their body temperature varies approximately with that of their environment. Consequently, the majority of snakes enjoy basking in warm sun to bring their body temperature up to a point where their metabolic rate is at an optimum level for mating, hunting and digesting their food. It is solely for this reason that snakes are more abundant in equatorial regions.

The ability to hibernate and aestivate are important where prevailing temperatures fall too low or rise too high for normal activity, and where food or water may be in short supply. To escape frost or excessive heat, snakes hide in burrows, beneath trees or under large rocks and enter a state of semi-torpidity. In this state, their metabolism is very low and they survive on the body fats built up during the previous active period. This semi-torpidity is also an important stimulus for breeding and can be induced to advantage in captivity. In desert regions, the temperature can fall below freezing at night and snakes either escape deep underground or rest under a large flat rock. Such rocks retain a certain amount of heat, which radiates slowly and maintains the snake's body temperature. This interaction is called thigmothermy (thigmo 'touch', therm 'heat').

Above: Elaphe obsoleta, *hatching.*

*Now the young snake has to fend for itself.*

## Reproduction

Courtship behaviour may be the result of several different stimuli. These range from increased activity, increased hours of sunlight (the photoperiod) or an increase in the food or water supply, to the release of pheromones ('scent hormones') by the female.

In only one species - the Malaysian tree snake, *Langaha* sp. - is there a clear difference between male and female, particularly in coloration. Otherwise, apart from size - the adult female is usually larger than the male - there are no sexual differences in snakes.

There is a strong tendency for males to outnumber females, but snakes are not as territorial as lizards and very few actually fight for the right to mate with a female. Once a fortunate male has secured a female, copulation either takes place immediately, as in garter snakes (*Thamnophis* sp.), or is preceded by a complicated courtship ritual, entailing dancing, biting, or body movement displays. Copulation can last from just a few minutes to several days, with the male inserting his copulatory organ, the hemipenes, into the female's cloaca and releasing jets of sperm. Following successful fertilization, the female undergoes a gestation period. Egg laying (oviparous) snakes deposit tough, leathery shelled eggs in a warm, moist location where the embryo can develop without drying out and at the end of the incubation period, perfectly formed snakes hatch out. Some snakes from colder temperate or high altitude regions have a longer gestation period, but give birth to live young. In their environment, ambient temperatures are too low for too long to permit successful incubation of the eggs.

Below: *The cavities, or pits, shown below the eye of this tree boa,* Corallus caninus, *are* sensitive to any shift in air temperatures, enabling them to detect warm-blooded prey.

# BUYING AND HANDLING

As with so many animals and plants today, wild snake populations are under a great deal of pressure. It is fortunate, therefore, that in many countries the snakes offered for sale or exchange are mainly captive-bred. As well as being more desirable from the conservation point of view, captive-bred animals are virtually disease-free, more easily coaxed to accept various foods and far more adaptable to an artificial environment. Of course, some species are so popular that captive-bred sources alone cannot meet the demand and wild specimens are also offered for sale. Although the pet trade does not pose a threat to native snake populations, Australia and certain South American countries, for example, have imposed a total ban on exporting these creatures. In Europe, overcollecting in the past virtually wiped out large populations of snakes and European species are now only occasionally offered.

## Buying snakes

Snakes are widely available, either from specialist dealers and private collectors or from the local pet store or tropical fish shop. The last two usually stock only the more common species, as their facilities tend to be limited. Do not disregard these outlets, however, as the specimens they offer, although few in number, are given individual attention. Their price may be slightly higher, but the result is usually a healthy snake. In general, shop assistants are very helpful, even if they lack specialized knowledge of snakes. On the other hand, they may be at a disadvantage if, for example, you are looking for a sexed pair of snakes.

As the demand for all herptiles grows, so do the number of specialist dealers. Some are only concerned with selling huge quantities of stock and many herptiles suffer in squalid, cramped and overcrowded conditions, but local and national animal welfare organizations do their best to keep these outlets to a minimum. Fortunately, most dealers are well organized, well stocked with a few specimens of a variety of different species and - most importantly - they clean out and monitor their display cages regularly.

A bona fide specialist dealer should also be able to give reliable information on sexing snakes and valuable tips on their captive requirements, food and maximum size.

Some dealers also operate a mail order service, but other outlets rely solely on mail order. They usually issue a periodic stock and price list, but you should confirm by telephone that the species you want is available. In many countries, it is illegal to despatch snakes via the postal service, so delivery is made by rail or a courier service. Unfortunately, this can be relatively expensive, especially where large snakes are concerned. Another drawback is that you cannot first inspect the snakes you are going to buy. It is your responsibility, therefore, to ensure that you buy from a reputable dealer. Although you can return poor quality specimens, you will be adding even more to the transport charges.

Specialist herpetological societies are an excellent source of first class snakes, both

**Above:** *The children's python,* Liasis childreni, *can attain 120cm(48in). It is a good-natured, robust and beautifully marked snake that clearly needs fairly spacious living quarters. Boas and pythons are more suited to hobbyists with experience.*

**Left:** *Snakes of the genus* Thamnophis *are perhaps the ideal beginner's species. They are small, easy to accommodate and very tolerant of life in captive conditions. Smaller snakes may live for 10 to 15 years and larger types as long as 25 to 30 years.*

common and unusual. They usually produce a newsletter with a 'for sale' section and fellow members may bring specimens to sell at meetings.

### Choosing a suitable species

Your first step should be to find out as much as you can about the species that interest you - their requirements, behaviour and maximum size. For instance, if you have limited space it would be careless to acquire a small python that will eventually require spacious living quarters. Beginners should clearly avoid a species that is difficult to feed or is known to have a nasty temperament. Most snakes can be kept singly or as pairs, but some species have a tendency towards cannibalism (king snakes, *Lampropeltis* spp., for example) and may require separate accommodation.

A far better prospect is to start with a species that is attractive, easy to breed and will thrive in captive conditions. There are many examples of such species and keeping them will give you confidence and a realistic idea of what is required to maintain the more 'difficult' snakes. Even after keeping 'easy' snakes, it would be thoughtless immediately to acquire a more demanding one. The secret of success is to be patient and gradually build up your experience and confidence.

Today's fluctuating prices mean that snakes are not cheap, so it is important to study a specimen before buying it. There are a number of points to consider before you make your final choice.

*Above:* Elaphe situla. *particular attention to* *Choose a healthy-* *the eyes, mouth, skin* *looking snake, paying* *and general alertness.*

**General appearance** Pay close attention to the alertness of the snake; when disturbed it should constantly taste the air with its forked tongue and try to avoid capture. The skin should have minimal blemishes and be unbroken, the mouth should be closed, with firm skin around the jaws, and the body should be free of lumps or other distortions. Overall, the snake should look healthy and adequately nourished, but not obese. Remember that very few dealers will agree to replace dead specimens once the 24-hour notification period has elapsed.

**Age and size** are generally related; the larger the snake, the older it is likely to be. Therefore, although snakes are relatively long-lived, it is wise to choose a smaller, younger specimen, which will live longer and, if wild-caught, adapt better to captivity.

If you require a particular sex or a sexed pair, then ensure that the dealer is capable of sexing snakes. If he is not, then ask him to agree - in writing - that you can return any incorrectly sexed snakes. Alternatively, learn the technique of differentiating between sexes using a probe (see page 120).

### Transporting snakes

To avoid undue stress when transporting a snake, place the specimen in a cotton 'snake bag', such as an old pillow case, making sure there are no holes that might form a possible escape route, and then secure the opening with a tight rubber band. Place the whole bag in a polystyrene box to keep the snake relatively warm. This is also the best way to send snakes by rail or other delivery service.

### Handling snakes

Never handle snakes more than is absolutely necessary, even if they are completely tame. The only time they need be held is when you are cleaning out the cage, transporting or inspecting specimens. On no account should you hold them during feeding, sloughing or when they are giving birth. Newly acquired snakes, particularly those from a wild-caught source, dislike being handled and will thrash about and bite repeatedly. Allowing them to remain partly wild is no bad thing, as they tend to behave more 'naturally' in this state. Remember that waving your hand in front of a snake causes unnecessary stress and incites aggressive behaviour and biting.

When handling juvenile or small snakes, grasp them gently but firmly to stop them wriggling. If the grip is too tight, you may cause internal bruising, which can prove fatal a few days later. The best method is to hold the neck between thumb and index finger, and allow the remainder of the body to rest

in the palm of your hand. Medium-sized snakes present few difficulties; grasp them behind the head with one hand so that the trunk rests in the other hand. Some snakes always try to bite, so apply a firm grip.

Larger snakes, over 1.5m(5ft) long, present a major handling problem. To prevent being bitten, wear leather gloves, taking care not to handle the snake roughly and remember that even seemingly harmless snakes can give a painful bite, often requiring a tetanus injection. Grasp the neck firmly and allow the rest of the body to curl round your other arm or, if the snake is a non-constrictor (for example a water snake, *Natrix* sp.) it can drape itself around your neck. Do not allow constrictors to wrap themselves around you, especially the larger pythons or boas, as they can easily suffocate you or they may come to harm when you struggle to detach them.

Take extra care when handling a gravid female snake, so that you do not cause egg breakages or stress. Hold her firmly round the neck and let her heavy body rest gently in the palm of your other hand. You may need help with large specimens.

Below: *Handle a small snake gently and carefully, letting it rest in your hands, rather than gripping it firmly, which will injure it.*

# HOUSING SNAKES

Providing appropriate accommodation for snakes is naturally important, but it need not be complicated; many 'difficult' species have been housed and bred in the simplest set-ups. The vivarium can vary from an all-glass aquarium to a stylish, custom-built wooden enclosure. In this section, we examine the housing options open to the snake-keeper.

## Small containers

It is far easier to maintain newly born or recently hatched snakes individually in a relatively small container. In this way, you will be able to keep a close eye on their health, food intake and subsequent growth. The same applies to smaller snake species, such as thread snakes, *Typhlops* spp. Suitable, easy-to-clean containers include thoroughly washed margarine tubs, plastic lunch boxes, plant propagating units or custom-made small plastic aquariums, depending on the size of the snake. The container must be well ventilated but escape-proof. If you opt for a plastic box, remove the lid and replace it with muslin, tightly secured with a rubber band. Alternatively, you can cut a large opening in the lid and cover it with aluminium mesh, as described in the section on housing lizards (page 22).

Provide suitable bedding material - such as crumpled newspaper, tissue paper, straw or chipped bark - a small ceramic water bowl and a hide. Burrowing snakes prefer a loose, fibrous substrate, such as chopped sphagnum moss of suitable depth. In a plant propagator or a plastic aquarium you could add rocks, branches and sturdy, small-growing live plants, such as ivies (*Hedera*) and various ferns.

It would be difficult - and unwise - to fit lamps into all but the larger plant propagators. A better prospect is to place a number of containers in a warm room or inside a heated vivarium. Many plant propagators already incorporate heating systems (some are thermostatically controlled). Position a fluorescent tube above the containers to provide light or place them near a well-lit window, but never position them in direct sunlight.

Above: *Plastic lunch boxes and similar containers make ideal starter homes for newly hatched or small snakes. It is possible to maintain a high level of humidity in them, and easy to keep a close eye on the snake.*

## Display vivariums

These are available in standard sizes or made to order in the shape and size of your choice. You can select either a normal glass aquarium or, better still, a glass-fronted wooden vivarium especially designed for reptiles. A glass aquarium is suitable for subterranean (burrowing) or semi-aquatic snakes, but the 'openness' can cause stress in shy species or a sore snout if the snake continually butts the glass in an attempt to get out. You can reduce stress by setting the aquarium into an alcove or by attaching black card to the glass sides and back.

Normal aquarium lids are not recommended for snakes, because they do not allow adequate ventilation and the lid is easily pushed open. Instead, buy a flat top from a herpetological supplier or construct your own, making sure there is sufficient ventilation and space for attaching lights.

## Building a vivarium

A ready-made vivarium, complete with lighting, heating and even humidity control, can prove rather expensive and you may decide to construct and assemble one yourself. The materials are not expensive and you can provide custom-built accommodation to suit your snake, your home and your pocket. The vivarium should be escape-proof, functional (i.e. equipped with heating, lighting and humidity controls) and easily accessible for cleaning. The size will depend upon the ultimate size of the snake to be housed in it and its docility. Most snakes remain hidden from view, curled up in a hide, and require relatively little space; arboreal species, however, need plenty of room to climb and explore.

Various materials are suitable for the construction, but wood is the cheapest and most versatile. Unprotected chipboard or plywood will absorb moisture, swell and warp, so waterproof them with sticky-back plastic or coat them with the waterproof paint used in concrete pool construction. Chipboard laminated with plastic or melamine is also suitable. Old cupboards or wardrobes can be adapted to make a vivarium and it is worth investigating furniture stores for imperfect cabinets that may only need a few adjustments.

When the basic construction is complete, you will need to drill holes to allow access for electric cables and openings for ventilation grills. Make sure you leave no gaps, as snakes are expert at escaping. Ventilation is critical; too small an opening will result in an accumulation of stale air, whereas a large hole creates a wind tunnel and the snake may catch pneumonia. Locate ventilation holes near the base of the vivarium, one on each side, so that stale air can be replaced quickly, and attach a strong metal gauze or custom-made plastic air vent to the outside.

Seal all joints and corners with the silicone sealant used in aquarium construction. This will reduce the risk of draughts and help to control the build-up of debris, which may

## Converting a cabinet into a vivarium

*Instead of building a vivarium from scratch, it is possible to convert an existing piece of furniture or buy a secondhand or slightly damaged item from a furniture store at reasonable prices.*

*With a few careful adjustments it can become an attractive vivarium, designed to fit in with the decor of your home.*

*Seal the vivarium base with waterproof sealant and check that there are no weak joints that might provide escape routes.*

*Make sure that the vivarium is standing on a sturdy base. Drawers should open smoothly, so as not to alarm the snakes.*

*You may need to remove the cupboard 'roof' to install the lighting equipment.*

*If you replace the side panels with glass, make sure that the vivarium is liberally stocked with plants to reduce stress on the snake. Plants increase the visual impact of the vivarium. Decor will depend on the species housed; live plants for small snakes, robust logs for larger species.*

*Siting the vivarium in an alcove will give the snakes an added sense of security and reduces stress. Bear in mind that you will need access to the vivarium for regular cleaning and maintenance.*

harbour bacteria and parasites. It also renders the vivarium leak-proof when it is washed down. Purpose-built trays made of wood, plastic or glass can be fitted inside the vivarium to hold the substrate and decor and are easily lifted out for cleaning.

Access to the vivarium can be via a wooden-framed glass door that swings out on hinges or by two sliding glass doors that move on plastic or wooden runners. Glass is better than perspex or acetate, because it is easier to clean and will not scratch. Make sure it has smooth, waxed edges to reduce resistance on the runners. Add a wedge or lock to the door to stop the snake pushing it open. Position lighting and heating sources towards one side of the vivarium and not centrally, particularly if you intend creating a 'hot-spot'. This will produce a gentle temperature gradient, enabling the snake to move to a cool spot once it is warm enough.

## Keeping the vivarium clean

Although snakes are generally undemanding reptiles, they do require clean comfortable surroundings in order to thrive. Inevitably, they will leave food uneaten, deposit faeces and occasionally upset water dishes. On a daily basis, therefore, remove any uneaten food and snake faeces, clean out the water bowl and provide fresh tap water. If a water bowl is tipped up, it may be necessary to clean out the entire vivarium. Where there are living plants in the vivarium, remove dead leaves regularly and examine the potting compost by scraping back the top layer to see if any unwelcome or potentially dangerous invertebrates and pests are present. As well as these daily tasks, it is important to clean out the vivarium more thoroughly, preferably each week or at least every fortnight. Disinfect the inside of the vivarium, together with all decor and snake hides, using a 3 percent solution of sodium hypochlorite (household bleach) or a disinfectant. Then rinse all the items very thoroughly with clean water and dry them. Discard all newspaper and thoroughly sterilize, rinse and dry all gravel or chipped bark substrate material. An organic substrate, such as chipped bark, should be completely replaced every two months or so.

Below: *Although* Nerodia rhombifera *is a water snake, this does not mean that it requires a totally aquatic set-up. Water snakes tend to develop sores and scale rot if kept constantly wet.*

Right: *A vine snake,* Oxybelis aeneus, *nestling in sphagnum moss. Live moss helps to maintain vivarium humidity - vital for skin sloughing. Give it light and a gentle spray with soft water.*

# HEATING AND LIGHTING

It is impossible to overemphasize the importance of providing the correct levels of heating and lighting if you wish to keep snakes successfully. The incorrect application of both heat and light is the single most common cause of stress-related illness in these reptiles, and can result in reduced activity, appetite and breeding capability, and increased vulnerability to disease. Snakes are ectothermic animals (see page 93) and each species has a certain temperature tolerance. For example, some temperate snake species regularly encounter low temperatures in their natural habitat, while certain desert species are quite accustomed to very high temperatures. In the wild, they are able to cope with these extremes by hibernating or aestivating. In captivity, it is much safer to avoid extreme temperatures, unless suitable retreats are available.

In a vivarium for snakes from warmer regions, separate sources of heat and light are essential. If you try to combine the two by, say, providing an incandescent lamp for light and as the only source of heat, it would mean leaving the lamp on constantly. Continual illumination is totally unnatural - and even harmful - and can result in visual disorders, abnormal behaviour and a breakdown in the breeding cycle. By manipulating day length (photoperiod) it is possible to influence the reproductive response in captive snakes.

## Heating the vivarium
Heating a vivarium can be considered under two separate headings: daytime and night-time. During the day, the majority of snakes enjoy prolonged basking and, ideally, the heat should radiate like the sun. To achieve this effect, use a spotlamp with a reflector that directs the heat and high illumination onto a limited area. Site the lamp to one side of the vivarium to create a thermal gradient, with high temperatures below the lamp and relatively low temperatures on the opposite side. A large flat rock directly below the lamp will distribute the heat uniformly and provide warmth for thigmothermic snakes. Once the snake's body temperature has reached the optimum level, it will be able to

pursue all its normal activities, such as breeding, hunting and digesting food.

If the vivarium is particularly large, you may require an additional heater to 'top up' temperatures. A ceramic infrared heater with reflector is ideal, as it gives out extra warmth but no light. Place it near the spotlamp, so that it does not interfere with the cool area in the vivarium.

Spot lamps and ceramic heaters can become very hot, so enclose them in a

Below: *A heating pad and a water heater. The water heater consists of a heated element enclosed in a glass tube. Once the water has reached a pre-set temperature, a built-in thermostat cuts off the power. This type of heater is essential for aquatic snakes that enjoy bathing.*

*The flat rectangular heating pad may also be equipped with a pre-set or adjustable thermostat to control the temperature. If not, install a separate thermostat. If the heating pad is to be used for an aquatic set-up, be sure to place it under the vivarium base and not inside.*

protective wire-mesh cover to prevent accidents; snakes are very inquisitive!

It is safe to assume that although temperatures at night will be slightly lower than during the day, even in tropical areas, you will still require a form of heating, and one that does not emit light. Night-time heat may come from an outside source, such as central heating or electric fan heaters, but bear in mind that snakes from different climates will require different temperatures. A better proposition is to provide heating in the vivarium in the form of a heater pad as used in wine making, or a horticultural soil-warming cable. Both are widely available in various wattages (some have variable controlled settings) and, as long as heat output does not exceed safe limits, they are very practical. They can be buried into a chipped bark substrate or placed under newspaper out of sight.

### Lighting the vivarium

Most of the commonly maintained snakes do not need special lighting and an incandescent spotlamp is quite adequate. Snakes are often secretive and many are also nocturnal, so they prefer subdued lighting and will not venture outside their hide in strong light. A vivarium for largely nocturnal snakes is best lit with very dull, red-orange coated bulbs.

Unlike lizards, very few snakes derive additional benefits from ultraviolet lighting (for example, Vitamin $D_3$ production in the skin), so elaborate lighting only enhances the overall appearance of the vivarium. The only snakes to benefit from UV lighting are predominantly diurnal and heliothermic (sun-loving) species, such as whips and racers, *Coluber* spp., various water snakes *Natrix* spp., and some rat snakes, *Elaphe* spp. The most commonly available lights that emit some UV light are 'natural daylight' fluorescent tubes and those designed for encouraging plant growth in aquariums. Remember that UV lighting of sufficient intensity is important if live plants are to be incorporated into a set-up: the taller the vivarium, the greater the intensity required. You can also buy high UV-output black lights, in the form of a tube or bulb (use BL or

Above: *Subdued or dark blue lighting during the day can encourage nocturnal snakes to emerge from their hides to hunt for* food. *The southeast European cat snake,* Telescopus fallax, *is a nocturnal species, but difficult to feed, as it eats only lizards.*

### Guide to preferred temperatures

| Type of vivarium | Daytime temperature range Min/Max | Night-time temperature range Min/Max |
|---|---|---|
| Cool temperate | 18-27°C(64-80°F) | 10-18°C(50-64°F) |
| Warm temperate | 21-32°C(70-90°F) | 16-21°C(61-70°F) |
| Sub-tropical | 27-32°C(80-90°F) | 21-24°C(70-75°F) |
| Tropical | 27-32°C(80-90°F) | 24-30°C(75-86°F) |
| Semi-desert/desert | 30-38°C(86-100°F) | 10-21°C(50-70°F) |

## Heating and lighting options

There are many timers on the market, and they vary in the functions they can perform. A basic model simply switches the lights on and off at preset times, and this is sufficient for most snake vivariums. Some models have a circuit breaker as an extra safety precaution.

The thermostat control box has three wires. One supplies power to the heater, the second is the power supply to the box and the third ends in a sensor, placed inside the vivarium and close to the heat source. Once the vivarium is warm enough, the power supply is cut off.

All the light bulbs featured here are available from most electrical stores. The normal incandescent bulb and the spotlamp are the most common. The blue spotlamp and the orange incandescent are not essential, but useful for naturally shy or nocturnal serpents.

Heater pads, such as those used for home brewing, are obtainable in various wattages. In a large vivarium, use several low-power heaters, rather than one high-power model that may become too hot to touch and is a potential fire hazard in the confines of a vivarium.

'black light' bulbs, not BLB 'black light bright' bulbs), which emit very little visible light. These are very effective when used in conjunction with a spotlamp or fluorescent tube to light a tall vivarium.

### Thermostats and time-switches

Because temperature is such an important breeding stimulus, the serious snake-keeper needs to be able to maintain strict control over it. A thermostat regulates the

temperature of the vivarium and can be used in conjunction with ceramic infrared heaters, heater pads or heating cable. Do not consider using a thermostat with an incandescent light that is also providing heat in the vivarium, since the constant switching on and off of the light can cause severe stress in snakes, being just as unnatural as continuous exposure to light and with the same consequences (see page 102). The thermostat should have a sensor so that when the vivarium reaches a

### Regulating body temperature

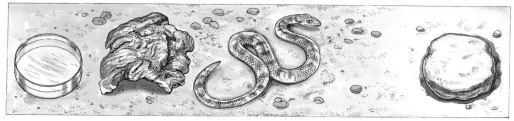

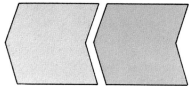

*Create a thermal gradient in the vivarium, so that the snake can move from one area to another, seeking the temperature best suited to its metabolic state.*

preset temperature the power to the heating appliance is cut off, thus ensuring that the vivarium and its occupants never overheat. Position the sensor near the heat source, not in a cool spot.

Although not as important a stimulus as temperature, photoperiod can also have a profound effect on snake courtship, especially in species from temperate regions, where seasonal fluctuations in day length differ quite considerably. A time-switch enables you to preset an exact illumination period and is indispensable if a vivarium is to be unattended for an extended period. Some of the more sophisticated models incorporate a dimmer (or rheostat), so that the light intensity increases or decreases gradually, as it would in nature. This dimmer

is also useful for snakes that experience seasonal fluctuations in light intensity.

**Installing electrical appliances**
It is vital that all lighting and heating equipment is safe and operating correctly before you add any stock to the vivarium. Conceal and insulate bare wires, ensure that plugs have the correct fuses, that cable is of a suitable electrical rating, that timers and thermostats are accurate and that all items are attached firmly in place. If you are uncertain about any aspects of the electrical installation, consult an experienced electrician. Many premature snake deaths have been caused by faulty or incorrectly wired electrical appliances. Inspect all appliances regularly for wear and damage.

### Guide to recommended photoperiods

| Type of vivarium | Average photoperiod (hours per day) | | | |
|---|---|---|---|---|
| | Spring | Summer | Autumn | Winter |
| Cool temperate | 12 | 14 | 12 | 8 |
| Warm temperate | 10 | 14 | 12 | 10 |
| Sub-tropical | 12 | 14 | 12 | 10 |
| Tropical | 14 | 14 | 14 | 12 |
| Semi-desert/desert | 12 | 14 | 12 | 10 |

# FURNISHING THE VIVARIUM

When it comes to furnishing the vivarium, you are faced with a number of alternatives and your choice will ultimately depend on the requirements of the species you wish to house. Bear in mind that even though certain species come from a similar habitat, their heating and lighting requirements may vary, so it is well worth finding out exactly where your snake originated and what its habits are, even if it has been captive-bred. (A more detailed analysis of individual species requirements is provided in the species section, starting on page 130.) In this chapter, we examine some standard vivarium arrangements that can be adjusted to suit individual species.

### The subterranean vivarium

This uncomplicated set-up is suitable for burrowing species, such as worm snakes, *Typhlops* spp. An all-glass aquarium is ideal,

Below: *A well-planted vivarium, with a water reservoir at the base, can look very natural and is pleasing to the eye. However, it takes time and effort to keep such a vivarium clean and maintain it in this good condition.*

as you can observe the snake burrowing. Line the aquarium base with 10-20cm(4-8in) of a loose, damp substrate material, such as shredded sphagnum moss, and place broken crocks or cork bark on top, along with a small water bowl. Be sure to change the substrate once a month.

The majority of burrowing snakes dislike intense heat and dry conditions and do not benefit from UV lighting. Place the vivarium in a warm room and heat it with a 40- or 60-watt incandescent lamp during the day. At night, a low-wattage soil cable buried deep in the substrate will provide sufficient heat. This arrangement will suit warm temperate species, such as the European worm snake, *Typhlops vermicularis*. Tropical burrowing snakes require slightly higher temperatures, so provide similar wattage lamps, along with a more powerful soil cable for them. It is important that soil cables are always connected via a thermostat, as *Typhlops* are very susceptible to drying out. In a centrally heated room, mist the vivarium frequently so that the damp moss does not dry out too quickly. A tight-fitting lid with adequate ventilation is essential.

## Substrate materials for the vivarium

*Pebbles*

*Chipped bark*

*Sphagnum moss*

*Vermiculite*

### The semi-aquatic vivarium

This habitat is mainly designed for water snakes of the genus *Natrix* and garter or ribbon snakes, *Thamnophis* sp., but it is not essential to keep these species in a semi-aquatic environment. If you do opt for this approach, however, an aquarium is a practical choice as there is little likelihood of leakage. There are two methods of furnishing a semi-aquatic vivarium, as explained below. In both cases, it is vital to provide a dry area; these snakes will not tolerate constantly wet skin, and may develop blisters and sores.

**1** Place a large water bowl in one corner, away from the heat source, and sink it into the substrate, which could be live sphagnum moss, chipped bark or even newspaper. Furnish the remainder of the vivarium with branches, a flat basking rock directly beneath the heat-lamp, and a suitable hide made from cork bark (available from florists).

Below: *The small rounded pebbles sold in aquarist shops are a popular substrate material. They look more natural than sheets of newspaper and are easily removed for cleaning. This set-up houses a yellow ratsnake,* E. obsoleta quadrivittata.

**2** Divide the aquarium into two equal parts with a piece of glass a few centimetres high. Secure the divider with aquarium sealant and ensure that the top edge has a smooth finish. Fill one half with water and the other with sphagnum moss or similar material. The hot spot should be in the dry area, which can be sparsely decorated with rocks, branches and a hide. If required, the water can be filtered and heated.

The water need not be deep in either type of vivarium, as snakes enjoy bathing in warm, shallow water. All the water should be replaced every time you clean out the vivarium, because shallow water will stagnate unless aeration and filtration are incorporated into the set-up. In this case, a partial water change every two to three weeks by siphoning is sufficient. You can do this using one of the siphonic devices available from aquarist shops. Remove as much detritus as possible and then top up the water level with ordinary tap water.

UV lighting is essential in a vivarium containing live plants, otherwise a spotlamp will suffice. Tropical water snakes may require more warmth, so consider adding a ceramic infrared heater, but do not place live plants close by. All spots and ceramic heaters can be controlled by a time-switch, and heater pads connected to thermostats are useful if extra heat is required at night. You may also wish to include a thermostatically controlled water heater (caption page 102).

Above: Lamprophis fuliginosa, *the nocturnal desert-dwelling house snake, rests under flat rocks heated by a spotlamp. It often drinks water droplets, so mist the vivarium at night.*

Left: *A flying snake,* Chrysopelea ornata, *hiding in a cavity of a gnarled log. Snakes are secretive creatures, preferring to spend a good deal of time out of sight. They need good cover to feel secure.*

### The terrestrial vivarium

The three vivariums described under this heading are very similar to those for lizards (see pages 34-38) and the same guidelines apply for setting them up.

**The arid vivarium**, as its name suggests, is suitable for snakes accustomed to barren desert areas, such as king snakes, *Lampropeltis* spp., and various other colubrids. Bark chippings are a suitable substrate material if you want an alternative to gravel, and the substrate, the decor and the inside of the vivarium should all be disinfected at least once a month. A removable tray at the bottom of the vivarium makes cleaning out easier. Do not forget to provide a small water container and a flat rock underneath the heat-lamp. Plant pots can be hidden behind rocks.

Although this vivarium is drier than the others, it should be lightly misted before you switch on the lights in the morning to imitate the effect of dew in the desert.

**The woodland or forest vivarium** can look most attractive and is suitable for a wide range of snakes from temperate, warm temperate and tropical regions. The overall humidity will depend on the species you keep, but remember that ambient temperatures should be more constant than in the desert vivarium. A ceramic infrared heater may be useful for those species that require more warmth. Most of the snakes that occupy this habitat enjoy frequent bathing, so provide a substantial water bowl. Rat snakes, *Elaphe*, whip snakes, *Coluber*, and smaller boids such as the rainbow boa, *Epicrates*

## Using plants to decorate the vivarium

*To achieve a natural look, enclose the vestigial bromeliad roots in Spanish or reindeer moss.*

*Secure the bromeliad by lightly binding the moss-covered roots to the log, using thin brown sewing cotton.*

*If preferred, you can 'plant' the bromeliad in a log. First, drill a suitable size hole to accommodate it.*

*Place the bromeliad in the hole and pack in peat or sphagnum moss to anchor it firmly in position.*

cenchria, enjoy climbing, and robust branches or even strong plants, such as the Swiss cheese plant (*Monstera deliciosa*), Philodendrons, castor oil plant, (*Fatsia*), mother-in-law's tongue (*Sansevieria*) and devil's ivy (*Scindapsus*) are all suitable for inclusion. Sink plant pots into the substrate or hide them behind rocks and logs, otherwise they will inevitably be displaced.

**The tall vivarium** is ideal for snakes that are almost entirely arboreal in their habits, preferring to explore, feed and rest among branches and higher vegetation. The overall dimensions depend on the species size but, generally speaking, the height of the vivarium should be three times the width. In other respects, the decor is similar to that described for lizards. Provide a large water bowl, some rocks and logs, and fix the hide in the top corner of the vivarium. A bird nesting box with a suitable opening is ideal.

Live plants are recommended only for smaller tree-dwellers, such as the beautiful longnosed tree snake, *Dryophis nasutus*, bronzebacks, *Dendralaphis* sp., and smaller rat snakes, *Elaphe* sp. Suitable plants include dragon trees (*Dracaena*), yucca, rubber plants (*Ficus*) and various bromeliads.

A vivarium for larger tree-snakes is best furnished with tall robust branches, as few house plants are strong enough to support species such as the emerald tree boa, *Corallus caninus*, or the green tree python, *Chondropython viridis*.

Lighting that provides some ultraviolet is essential, not only because most of the species suited to this vivarium enjoy extended periods of basking, but also because any plants included in the vivarium will benefit from it. In a normal forest canopy, light diminishes rapidly towards the ground and the same effect works well in a tall vivarium. Towards the top, the illumination can be quite high, so position a UV spotlamp so that it shines onto a wide branch or log, thus providing the snake with both heat and UV radiation. The power of the spotlamp depends on the species and the size of the vivarium.

A heater pad and a thermostat can be used as a 'back-up' to maintain the temperature levels, but if live plants are present, ensure that the soil does not dry out too rapidly.

Above: *Plants are unsuitable in a tall vivarium designed to house the rainbow boa,* Epicrates cenchria, *or other bulky arboreal snakes. Instead, use strong twisted logs.*

Below: Tillandsia *species are suitable for most dry vivariums. Position them away from the heat source, provide UV light and spray them gently from time to time.*

**The simple vivarium**

A simple hygienic vivarium is ideal for hobbyists who either have limited time to spare for cleaning out the vivarium or who keep large numbers of snakes in individual containers. Fortunately, most snake species adapt well to the arrangement and the lack of furnishing allows you to view the occupants more easily.

The ideal substrate is newspaper, which quickly absorbs snake wastes and is cheap and easy to replace. The remaining decoration can be sparse; a flat rock underneath the heat source, plus a branch or two, a suitable hide and a large water bowl. Disinfect and thoroughly rinse out the whole vivarium and its furnishings every week.

Most heating and lighting arrangements can be incorporated into this set-up, the only limitations being the species of snake and the size of the vivarium. For most snakes, a spotlamp during the day and a heater pad during the night are totally acceptable. Place the pad below several thicknesses of newspaper and locate the spotlamp so that it illuminates a flat rock or log.

# FEEDING SNAKES

Snakes are generally considered to be one of the wonders of nature when it comes to feeding. Their methods of capturing, subduing, killing and swallowing prey are second to none, while their digestive system is one of the most powerful known. A snake detects its prey by smell/taste (using its Jacobson's organ, page 91), by sight, by detecting changes in temperature (page 92) and by sensing vibrations. If a potential victim has an unfamiliar scent or appears too large, then the snake will adopt a defensive behaviour or flee.

Because snakes are totally carnivorous, they do not expect - or require - the same variety of food as lizards and other herptiles. Snakes generally favour one or two particular foods; for example, rat snakes are predominantly rodent eaters. However, if other foods become abundant or their normal prey becomes scarce, they will adjust accordingly, and rat snakes will also eat birds, lizards or other snakes.

Although snakes seem to be attracted by moving prey, they will also take dead food if the scent is correct. In captivity, scent can be used to advantage, especially when a particular food source is scarce. For example, rubbing an unfamiliar food against the normal food transfers part of the scent and can persuade the snake to eat it. In fact, an Indian python once attempted to swallow a rubber ball that had been rubbed against a mouse! If food is in short supply, low-fat, fresh meat dusted with a multivitamin supplement, can be used as an alternative. However, some species or individuals are very choosy and will not accept alternatives.

Snakes derive all their nutritional requirements from a single food, basically because the animals they eat are mainly vertebrates and contain all the vitamins and essential elements that they need. Insectivorous snakes prey on a variety of insects and other invertebrates.

Here we look at the various commonly available foods, plus a few specialist diets.

**Insects and other invertebrates (except earthworms)**
Some snakes are so small that invertebrates are the only food small enough for them to ingest. Green snakes, *Opheodrys*, and black-headed snakes, *Tantilla*, from North America specialize in spiders, crickets, grasshoppers and caterpillars. Ringneck snakes, *Diadophis*, do not specialize in insects, but do supplement their normal diet by taking soft-bodied insects. Many recently hatched or newly born colubrids take insects for the first few months, particularly whip snakes, *Coluber*, which are extremely small to begin with (see also page 142).

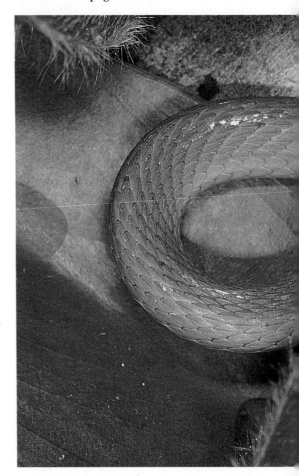

Right: *The North American rough green snake,* Opheodrys aestivus, *is a small serpent that feeds mainly on suitably tiny insects, such as crickets and spiders.*

**Methods of subduing prey**

| Species | Method |
|---|---|
| Garter snakes (*Thamnophis*)<br>Green snakes (*Opheodrys*)<br>Water snakes (*Natrix*) | Straightforward 'grab and swallow.' |
| King snakes (*Lampropeltis*)<br>Rat snakes (*Elaphe*)<br>Whip snakes (*Coluber*)<br>Bull snakes (*Pituophis*)<br>Boas and pythons | Small prey: 'grab and swallow'. Larger prey: usually subdued by constriction. |
| Montpellier snake (*Malpolon*)<br>Tree snakes (*Dryophis/Oxybelis/Chrysopelea*)<br>Egg-eating snake (*Dasypeltis*)<br>Night snake (*Hypsiglena*) | Several techniques, depending on size of prey, including 'grab and swallow', constriction, and by using weak venom via back-fangs. |

It is easy to collect insects, spiders and grubs from under rocks or by sweeping a net through a grassy meadow during the milder parts of the year. Crickets or locust hoppers are available from specialist food laboratories throughout the year and are not expensive to buy. Keep them in an old aquarium containing crumpled newspaper and place them in a warm light location. They feed on grasses, lettuce and carrot. (If you want to breed these creatures, see *Feeding lizards* page 42 for further information.)

### Earthworms

Earthworms are an essential nutritional source for a variety of serpents, including Dekay's snake, *Storeria*, worm snakes, *Typhlops*, the young of many water snakes, especially dice snakes, *Natrix tessellata*, and viperine snakes, *N. maura*, and juvenile to subadult garter snakes, *Thamnophis*. As earthworms are relatively low in some essential elements, such as calcium, it is a good idea to give them a light dusting with a multivitamin powder.

Worms are easy to collect in bulk by

digging in damp, airy soil or among rotting leaves; the large night-crawlers found on lawns after heavy rainfall make a sizable meal for adult garter snakes, for example. Alternatively you can collect some earthworms and breed them in an old aquarium. Put equal portions of earth, humus and rotting leaves in the aquarium and position it in a cool, shady location where the substrate will not dry out. Place flat rocks or rotting logs on the soil surface and the earthworms will regularly lay hundreds of eggs there. These can be hatched and reared in the aquarium or divided among several containers.

### Fish

Fish form the main diet of many water and garter snakes. Fishmongers can supply fresh, unfilleted whitebait, which should be cut into suitably sized strips, complete with all bones and scales. Similar-sized fish are abundant in shallow pools and slow-moving streams and can be caught by rapidly sweeping a net through the water. Avoid using fish such as sticklebacks with sharp spines that will inevitably cause mouth injuries or internal damage leading to infection. If other fish become unavailable, then goldfish are easy to acquire all year round, but they are a relatively expensive food.

Below: *Subterranean worm snakes, such as* Typhlops lalandei, *hunt for and capture earthworms by sensing their vibrations, by smell or simply by bumping into them!*

### Amphibians and lizards

In their natural habitat, some snakes feed almost exclusively on amphibians, lizards or even other snakes. To provide this diet in captivity could prove difficult, expensive or both. King snakes, *Lampropeltis*, are well known for their snake-eating habits, but can generally be coaxed into sampling small rodents. The same applies to the European lizard-eating rat snakes. However, the toad-eating Eastern hognose, *Heterodon platyrhinos*, cannot be persuaded to eat anything else. If you are still keen to maintain these snakes, reptile dealers may be able to provide suitable amphibians or cheaper lizards, such as house geckos, *Hemidactylus frenatus*, or you could attempt to breed your own.

### Birds

Many snake species will accept birds in their diet and they have a high nutritional value. Day-old chicks can be obtained frozen from

Above: Thamnophis elegans *eating a frog. In some countries, wild amphibians are protected, so try to breed your own supply.*

specialist suppliers and are not expensive if bought in quantities of a hundred. However, as they are low in calcium, dust them with a mineral supplement and do not offer them as a sole food item. A single chick is ideal for baby bird-eating snakes, whereas a green tree python, *Chondropython viridis*, will require up to a dozen at a time. Large ratsnakes, most pythons and boas devour large birds, such as pigeons, chickens or bantams.

### Rodents

Rodents form the major food for serpents and, fortunately, are easily obtained throughout the year or bred in the home. Day-old pink mice and rats (called 'pinkies') are probably the commonest food for juvenile rodent eaters and are available frozen and in

**Guide to treatment of vitamin deficiency**

| Vitamin | Deficiency symptoms | Treatment |
|---------|---------------------|-----------|
| A | Loss of appetite, impaired vision, kidney complications, poor epidermal development and decreased resistance to disease. | Add cod-liver oil to diet or force feed. Injections of vitamin A may be necessary. |
| B | In fish-eating snakes, an excess of the enzyme thiaminase can destroy the vitamin $B_1$, causing convulsions and slight or total paralysis. | Add liver extract to diet. Severe cases may need intramuscular injections of a thiamine drug. |
| D | Vitamin $D_3$ is essential to increase the absorption of calcium and phosphorous in the small intestine. Without it, inadequate amounts of calcium are deposited in the bones, causing rickets, poor skeletal development or deformity. | Add fish-oil to diet or syringe an egg yolk/ multivitamin solution down the snake's throat. Provide sufficient levels of calcium in the diet. |

bulk from specialist breeders. Alternatively, your local pet shop may be able to provide a limited number of pinkies, although you must be discreet when buying them in front of other customers. Adult mice, rats, gerbils, hamsters, guinea pigs and rabbits are also available, either freshly dead or frozen from specialist suppliers or, sometimes from a local pet shop. Occasionally, freshly killed rabbits and hares are sold in a traditional market and these are suitable for pythons.

If you keep a large number of snakes, the annual cost of buying rodents to feed them can become quite high and you may wish to breed your own. Mice and rats require little space and attention and their reproductive output is very high. Larger rodents are equally fertile, but may require more space. Nevertheless, the cost of breeding rodents for food will be much lower than buying them.

There is one final point to consider concerning animal foods. Do not offer living birds or rodents as food, since they can deliver a painful - even fatal - bite, especially if they are frightened. In any case it is illegal in many countries to offer live prey. This means that you will have to find a humane method of killing them first. Fish, frogs, rodents and birds can be swiftly killed by a quick but powerful blow to the head. Feed the animals immediately to the snake or refrigerate them for future use. In fact, freezing is essential if you are breeding large numbers of pinkies.

**Vitamin supplements**
Although a normal diet should contain the correct balance of vitamins, minerals and trace elements, it is sometimes necessary to add a supplement. The most critical times are during - and immediately after - illness or gravidity. Multivitamin supplements before and after hibernation or aestivation and before the reproductive season can be beneficial. These additives can also help to promote a healthy growth rate and sound skeletal and epidermal development in insectivorous and juvenile snakes. Multivitamin powders and solutions are available from all pet stores, and some are specially formulated for small animals, including herptiles. The products should be administered by dusting them liberally over the food or dissolving three or four drops in the drinking water.

Left: *A small rodent is a suitable meal for this* Lampropeltis getulus californiae. *It also accepts birds. House kingsnakes separately in captivity to avoid the risk of cannibalism.*

Below: *When they first hatch, colubrids tend to be quite small and weak. Offer them small crickets or similar sized insects, liberally dusted with vitamin supplement.*

Above: Dasypeltis scabra *eating an egg.* Dasypeltis *species are one of the marvels of the serpent world. The fact that they are able to dislocate and stretch their jaws over an egg four times the diameter of their head and gain all their nutritional requirements from this one food is remarkable. Juvenile or sub-adult specimens may require small quail eggs.*

**When to feed snakes**

In captivity, you should offer the snake its normal foods only when it is hungry. If the snake refuses them, offer alternatives until you find an acceptable one. Should the snake steadfastly refuse to eat, then either wait a few more weeks before offering the same food again or you could try force feeding it (see page 119).

Even when it has just been fed, a snake may still appear hungry, searching for more food among the rocks and logs. (A satisfied snake will hide under a rock or bask under a spotlamp.) Resist the temptation to offer more food except, perhaps, in the case of highly active, young specimens. An obese, overfed snake is unsightly and unhealthy and runs a higher risk of digestive or heart complications. As a guide, a healthy adult corn snake (*Elaphe guttata*) will need one mouse or rat every six to nine days, whereas a bantam will satisfy a 150cm(5ft) python for up to two weeks.

A newly acquired snake will rarely eat during the first few days in a new home. This is nothing to be alarmed about; once it has settled down, it will begin eating greedily. A few species are notoriously difficult to feed; indeed, some larger pythons have been known not to eat for a year in new captive conditions. Even the occasional specimen from an otherwise 'easy' species may refuse food for a lengthy period for a number of reasons, including disease, gravidity or stress. Snakes are very adept at fasting, as their metabolism is much slower than that of other higher vertebrates. However, once a snake begins to appear emaciated, it is time to consider force feeding.

## Force feeding

There are times when force feeding becomes a necessity - when normal foods are in short supply, for example, or when rearing newly hatched or difficult and choosy snakes. The main problem lies in forcing open the snake's jaws without causing undue stress or breaking the delicate teeth or jawbones. It is natural to approach the task with some trepidation, especially as the snake may resist having an object thrust into its gullet, but with a gentle approach and patience, force feeding becomes part of a snake-keeper's life.

Start by selecting a suitably flat, blunt instrument in proportion to the size of the snake's head. A matchstick would be suitable for babies, a lollipop stick or tongue depressor for medium-sized snakes and a spatula for pythons or boas. Moisten it before attempting to lever open the snake's mouth and make sure that the food is to hand and also moist. Never force the lever into the snake's mouth, as this will cause damage. Instead, gently oscillate the lever against the clenched teeth until the jaws relax and open. At this point, you can introduce the food into the gullet opening and, again, gently oscillate it. With small snakes this may be quite a delicate operation, but because they are so small, only one person can carry it out. Larger snakes are best restrained by one person and force-fed by another. Once the food reaches the gullet opening, the snake should begin to swallow, although the food is often regurgitated later.

Another method of force-feeding is to use a hypodermic syringe with a plastic tube instead of a needle. Gently push the tube down the snake's gullet and slowly inject the liquid, which might consist of raw egg yolk and a multivitamin supplement. This is less stressful than conventional force-feeding, but there is a strong risk of overfeeding, especially in young snakes. Seek help if in doubt. Snakes that are force-fed should eventually begin to eat normally, once they are used to the food or have sufficient energy to capture their own prey.

# BREEDING SNAKES

Breeding snakes is the ultimate objective for most hobbyists. Not only is it a fascinating aspect of their maintenance, but also a measure of how happy the snakes are in captivity; they will not breed if climatic conditions are incorrect, if the vivarium is dirty or overcrowded, if there is any evidence of bullying or if the snakes are over- or underfed, diseased or too old.

Until quite recently, the captive breeding requirements of snakes in general were greatly misunderstood and there was too much emphasis on attempting to simulate their natural habitat and associated climatic conditions within the vivarium. Although this approach may be necessary for a limited number of the more difficult species, most snakes can be bred under the controlled - and sometimes unfamiliar - conditions within a simple 'hygienic' set-up (see page 111).

Some species, such as water snakes, are very easy to breed - usually it is just a case of introducing a male to a female - but other snakes may require specific stimuli.

## Sexing snakes

In order to breed snakes, you must first acquire a sexed pair. Beginners or inexperienced snake-keepers may have difficulty in establishing the sex of their snakes, but a specialist supplier should be able to sell you specific sexes. However, this is not always the case and it is useful to know how to differentiate between them. Juvenile snakes are difficult to sex because of their small size; when fully grown, female snakes are generally larger than males.

Apart from a few species, there are no marked differences in colour that make it easy to identify male and female snakes. However, in most adult males the tail (measured from the cloaca) is proportionally longer than in the female and noticeably swollen near the cloaca. The hemipenes are situated here in an inverted position and only emerge during copulation. If judging tail length is difficult, then you may have to use a probe. This entails inserting a smooth, round-ended glass rod, lightly lubricated with petroleum jelly, into the cloaca, towards the

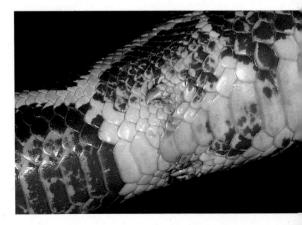

Above: *In primitive snakes, such as the* Boa constrictor, *the male spurs - remains of external limbs - are visible near the cloaca.*

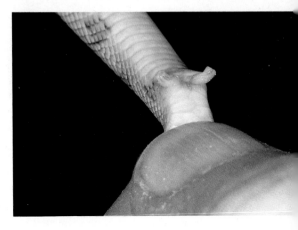

Above: *One way to sex snakes is to gently squeeze the cloacal region. In males, this will cause the hemipenes to 'pop' out.*

tail end. (The diameter of the probe depends on the size of the snake's cloacal opening.) If the probe travels some distance, then the snake is a male, whereas if there is little movement, it is a female. It is absolutely vital that this procedure is carried out with the utmost gentleness and patience, otherwise you risk incurring injury and stress or maybe even ruining the snake's breeding potential.

If you are unsure, consult an experienced snake-handler or join a herpetological society, where other members may be able to help.

## Conditioning for breeding

If snakes are to breed, they must be active, feeding regularly and free from illness, injury or disease. Undernourished snakes will be too weak to embark on courtship, and their hormone levels will be inadequate. Males will have a low sperm count and females will not ovulate (i.e. release eggs).

Even healthy snakes may only breed under certain conditions or as a result of particular stimuli, depending on their natural habitat.

**1** Snakes from cool temperate, temperate and mountainous regions have a defined breeding season, governed by ambient air temperatures and light intensity. In captivity, these snakes need a period of hibernation, so gradually decrease the light intensity and day length (photoperiod) and place them in an escape-proof aquarium filled with a dry material, such as shredded sphagnum moss or straw. Put the aquarium in a cool, frost-free spot at a temperature of 7°C(45°F) - an unheated attic or garage is ideal - and leave it for 7-12 weeks. All the snakes to be hibernated must have sufficient reserves of body fat and should not be fed immediately before dormancy.

As they warm up, offer them an immediate supply of food and gradually increase the intensity of light and extend the day length. Courtship should begin three to six weeks later. The advantage of breeding snakes in captivity is that you can induce a 'breeding season' at any time of the year, by substituting a fridge (not a freezer) for the cool garage or attic.

Below: *Probing a snake requires extreme care. Only use the correct instruments and be sure to seek help if you are uncertain.*

### Sexing a snake

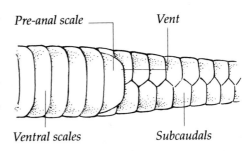

Pre-anal scale — Vent

Ventral scales — Subcaudals

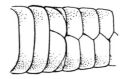

*The vent is situated by the large pre-anal scale (shown at top). In some species, this is divided in two (left).*

### Using a probe

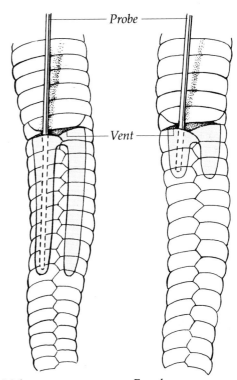

— Probe —

— Vent —

*Male*

*Female*

In the male snake (above left) the probe will travel some distance (about 10 subcaudal scales) down the hollow

tubular inverted hemipenis. In females, (above right) it will only travel a short way (about 3 subcaudals). Never push too hard.

121

2 In snakes from subtropical and tropical regions, breeding can take place throughout the year and is determined by other factors, including an increase in humidity, rainfall or food. An effective strategy with captive snakes is to decrease the vivarium temperature by 4-6⁰C(7-12⁰F) for several weeks. Such a drop in temperature rarely occurs in tropical areas and is unfamiliar to the snakes. When you raise the temperature again, the sudden increase in warmth should stimulate breeding.

3 Desert snakes can breed at any time of the year but are more likely to do so after a period of aestivation. In the wild, aestivation occurs when the air temperature reaches unbearable levels and rain or humidity is very low. The night-time temperatures are only affected in the sense that there is no cloud cover, which means temperatures will fall to freezing point. Aestivation ends when daytime temperatures fall due to clouds and rain. If there is cloud cover at night then temperatures will be higher than when drought occurs.

Most desert snakes do not need such stimulation and breed regularly. However, the following procedure may encourage desert snakes that refuse to breed, but only if sufficient aestivation quarters can be

Above: *The bull snake*, Pituophis melanoleucus sayi, *requires a long period of hibernation before it is ready to breed.*

integrated into the set-up. For example, they will need a fairly deep burrow, which should be cooler than the ambient air temperature.

First, cut down daily misting of the vivarium to reduce humidity, and then increase temperatures in the day to 35⁰C(95⁰F),but not above 38⁰C(100⁰F). Reduce the night-time temperature slightly to around 18⁰C(64⁰F); this may be difficult in summer unless some sort of fan cooler can be fitted into the vivarium. Six to eight weeks later, gradually decrease daytime temperatures on a daily basis, increase mistings to three or four times a day and increase the night-time temperature to 18-21⁰C(64-70⁰F) maximum. This should encourage activity and, after feeding and sloughing, the snakes should start to mate.

In all snakes, other stimuli may play some part in reproduction, particularly the discharge of natural sexual odours from the female, known as pheromones. After a period of dormancy, both snakes will slough their skin and at this time the pheromones are at their strongest, driving male snakes into a wild frenzy - famous in garter snakes - in an attempt to mate with a female.

## Mating and frequency of breeding

Generally speaking, it is not necessary for one sex to outnumber another in order for breeding to be successful. However, in garter snakes, for example, the presence of other males will lead to a vigorous and successful mating, because usually the stronger, but not necessarily the biggest, male will 'win' the female. A further advantage is that in future breeding programmes a different male can mate with the same female, thus producing genetically different offspring. The first generation can then be interbred, either with themselves or with the male that was not their parent, with less risk of producing a deformed second generation. Obviously, after the second generation you must introduce a new male or female to invigorate the breeding line.

When a pair of snakes are ready to breed they may either mate immediately or perform intricate courtship dances, rub bodies and bite each other. The male then inserts one of his hemipenes into the female's cloaca and releases small sperm sacs. (In some species, the hemipenes are covered with minute bristles to ensure a firm hold.) The union can last from a few minutes to several days, but once completed the eggs are either fertilized immediately or the sperm is retained and stored for future use.

Most temperate snakes produce only one or two broods per year, following a prolonged mating period during spring and early summer. To produce young throughout the year would be fatal, as their survival is entirely dependent on the availability of food and suitable climatic conditions only encountered during a fairly limited period in the wild. Even given ideal conditions and adequate food, these snakes will not breed throughout the year. By contrast, snakes from warmer regions (but not desert conditions) will either produce a large clutch of eggs once a year (as in pythons and boas) or smaller clutches three or more times a year (as in various tropical colubrids). In the wild, they are accustomed to regular supplies of food and water and ideal climatic conditions all year round.

## Gravidity

Following a successful pairing, the female eventually begins a period of gestation, or

Above: *A pair of corn snakes,* Elaphe guttata, *mating. Some* *snakes take part in an elaborate dance ritual before mating.*

'gravidity'. The vast majority are egg laying (oviparous), with the major part of the embryo development occuring within the egg after it has been deposited. In many boas and some smaller colubrids, such as garter snakes, the eggs are retained and develop within the females (a type of livebearing known as ovoviviparity). The yolk-sacs nourish the embryos during a considerably longer gestation period, after which perfectly formed young are deposited covered in a thin membrane. They quickly break free of this to lead an independent, if initially vulnerable, existence.

It is essential to determine whether a female is gravid, as she will require close attention. Initially, gravidity can be difficult to detect, although the female may become aggressive towards other snakes or when handled. Later, she becomes more rotund and basks under a warm lamp for long periods. By allowing her to move gently through your hand, you may be able to detect lumps, although this may not be the case with livebearers. Once you are sure the female is

## Guide to gestation periods and brood size in livebearing snakes

| Species | Gestation period in days | Number of young |
|---|---|---|
| Red-bellied snake (*Storeria occipitomaculata*) | 80 - 120 | 10 - 23 |
| Garter snakes (*Thamnophis* sp.) | 100 - 150 | 8 - 75 |
| Turkish sand boa (*Eryx jaculus*) | 130 - 175 | 6 - 21 |
| Emerald tree boa (*Corallus caninus*) | 172 - 210 | 2 - 12 |
| Common boa (*Boa constrictor*) | 115 - 240 | 11 - 52 |

gravid, isolate her in warm, humid, clean quarters and offer food regularly.

All snake eggshells are perforated by millions of microscopic holes that allow moisture and respiratory gases to permeate freely. A box filled with loose, moistened sand or sphagnum moss is a suitable egg container. Livebearers also prefer to deposit their young onto such a substrate.

Once deposited, the snake usually gives the eggs no further attention, although some pythons and colubrids coil their bodies around the clutch to maintain a slightly higher incubation temperature. However, the eggs and young of other species should be removed as soon as possible, so that they are not crushed or eaten.

### Incubating eggs

Eggs should always be removed carefully and never turned, otherwise the yolk-sac may cover and suffocate the developing embryo. Place the eggs in a warm, humid and sterile environment, such as a plastic food container, half-filled with sand, perlite or vermiculite. The substrate should be uniformly damp, but not soaking wet, and the eggs should be buried so that only the top third is visible. Make a few small ventilation holes in the lid and secure it tightly to maintain humidity within the container. Using this method, the eggs will require only

an occasional light misting to retain sufficient moisture during incubation.

The next step is to provide warmth. Place the eggs in an airing cupboard, a heated vivarium, a heated plant propagator or store them on a heater pad. Make sure that the temperature remains sufficiently high. All snake eggs should be incubated at about 26-32°C(79-90°F) during the day, with a slight drop at night. A thermostat is very useful if the eggs are incubated in a heated vivarium.

During incubation, the eggs may become slightly discoloured or shrivel at a later stage, even if sufficient moisture is present. This is the result of the embryo absorbing the yolk-sac and is quite normal. Microorganisms thrive in this humid atmosphere and if a fungus attacks the eggshell, gently wipe the eggs and move them to another container with fresh substrate. Eggs that become mouldy and completely discoloured are probably infertile; unfortunately, quite a high percentage of eggs may be infertile but, again, this is quite natural.

### Hatching and rearing

If the eggs show no signs of hatching after about 150 days (200 days for pythons), they are almost certainly infertile or the embryo has died. Fertile eggs will begin to hatch out, a process that may take several hours. Hatchlings use a sharp projection on the

snout called an egg-tooth to slice into the leathery shell, and after they have cut a suitable exit, the snakes remain in the shell to absorb the remainder of the yolk-sac.

Place newly emerged or newborn snakes immediately into small individual containers and keep them warm. The young of smaller snakes tend to be delicate, but ratsnakes or boas are relatively large. The first thing the baby snake will do is to shed its skin, so provide ample humidity and a small rock in the container to aid sloughing. The diet of the young varies from snake to snake, but it may be difficult to acquire suitably sized food for some of the smaller species. The best approach is to offer a variety of foods or try the 'rubbing' method on different foods (see page 112). Eventually, the young snake will become so hungry that it must eat. Add a multivitamin supplement to all food and water to promote strong growth. Baby snakes grow quickly; garter snakes should reach maturity within 15 months, but some boids may take four years. Do not allow juveniles to hibernate or aestivate for the first year, so that they attain sexual maturity more swiftly.

Above: *A milksnake,* Lampropeltis triangulum sinaloae, *laying a clutch of eggs.*

*Remove the eggs to a warm sterile container soon after laying for a period of incubation.*

## Guide to clutch size and incubation periods in egg laying snakes

| Species | Clutch size | Incubation period in days at 30°C (86°F) |
| --- | --- | --- |
| European leopard snake (*Elaphe situla*) | 3 - 8 | 58 - 65 |
| Corn snake (*Elaphe guttata*) | 10 - 30 | 55 - 60 |
| Grass snake (*Natrix natrix*) | 12 - 45 | 45 - 55 |
| Common king snake (*Lampropeltis getulus*) | 5 - 24 | 42 - 48 |
| Green tree python (*Chondropython viridis*) | 12 - 30 | 47 - 56 |
| Indian python (*Python molurus*) | 20 - 60 | 58 - 68 |
| Reticulated python (*Python reticulatus*) | 15 - 110 | 70 - 85 |

# HEALTH CARE

Keeping your snakes in good health is clearly the top priority. In this section, we consider the value of initial quarantining for new acquisitions and then look at the most commonly encountered disorders and how to diagnose and treat them. If you are uncertain, consult a veterinarian, who may in turn refer you to someone with more specialized knowledge of reptile disorders. At meetings of a herpetological society you can learn from the experience of other enthusiasts. A snake pathologist may well be one of the members.

### Health check and quarantine

Snakes are relatively expensive to buy and you should carefully examine new acquisitions for signs of illness or disease. Even apparently healthy specimens may succumb to illness as a result of stress during transportation or undernourishment.

Do not introduce a new snake immediately to an established collection. Remember that even in a separate enclosure one snake can pass disease on to others nearby. Give all new acquisitions a period of isolation, or quarantine, well away from other snakes. The quarantine enclosure can be quite simple - an old aquarium is ideal - but make sure it is clean, appropriately lit and warm. Decoration can be limited to a newspaper base, a few rocks or logs, a water bowl and a hide. Keep the snake in these separate quarters for at least two weeks, and frequently replace the bedding, food and water.

Above: Ophionyssus natricis, *a snake mite, (x120) from a ratsnake.* *Notice the piercing mouthpart inbetween the forelimbs.*

### Parasites

There are two groups of parasites; the external ectoparasites that live on the skin and have piercing mouthparts to draw up the host's blood, and internal endoparasites. During the quarantine period, ask a veterinarian to examine the snake's faeces regularly to ensure that any intestinal worms or a bacterium, such as *Salmonella*, can be treated quickly.

**Snake mite** (*Ophionyssus*) is the most familiar external parasite to attack snakes. These tiny, rounded black creatures congregate particularly around the eyes or other soft areas. Small infestations need not cause concern, because snakes seem resistant to

Left: *Cellulitis, an infection of the skin tissues, gives this black ratsnake a swollen appearance. Low humidity or poor hygiene also caused the retained spectacle.*

Right: *Belly scale rot in a Florida kingsnake. The condition, caused by an infection, can lead to scale loss and death if not treated.*

them and they are very easy to eliminate. However, wild snakes may play host to mites carrying potentially dangerous blood diseases, which can infect your whole collection. A dichlorvos strip is an effective means of eliminating mites. Place a small strip in the top corner of the vivarium for about 36 hours and repeat the treatment every 10 days until the problem has completely cleared up. Never leave the strip permanently in the vivarium, as it can produce dangerous side-effects. If dichlorvos treatment is ineffective, try bathing the snake for an hour in a 0.2 percent solution of dichlorvos and trichlorvos (available from chemists), keeping the snake's head out of the solution.

**Ticks** (Ixodidae) are larger than mites and easy to see, being swollen and red (from the snake host's blood). They are potential carriers of blood disease. Unfortunately, it is difficult to remove them alive, because of their hooked mouthparts, which can cause localized infection if left embedded in the snake. The best method is to wipe the snake down with a 0.4 percent solution of dichlorvos and trichlorvos or, for heavy infestations, a methylated spirit solution. This suffocates the ticks, which can then be carefully unhooked with tweezers by turning them 'head over heels'. During such treatment, you can also place a dichlorvos strip in the vivarium as described above.

**Nematodes** and **trematodes** are intestinal worms and flukes that survive in the intestinal tract. Heavy infestations cause

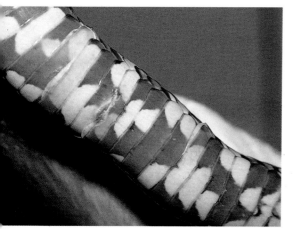

emaciation, loss of appetite or infection if they penetrate the intestinal wall. Sometimes, they move around the alimentary canal and form cysts, causing blockages or disfiguring lumps. They can be eliminated by force feeding the snake with an aqueous solution of mebendazole or piperazine citrate. Be sure to consult a veterinarian, who will prescribe the required strength.

**Tapeworms** are more serious intestinal worms. Even small infestations are liable to result in severe undernourishment, emaciation or internal bleeding and infection caused by the worms' hooks and spines. Treat the infestation with a course of praziquantel, as prescribed by a veterinarian.

**Blood parasites** are deadly internal parasites that can be introduced via mites, ticks, open skin wounds or contact with other reptiles. They need swift treatment, otherwise the snake may die within 40 days. One of the most pernicious blood parasites is *Entamoeba invadens*, a single-celled protozoa that causes gastro-enteritis, producing symptoms that include include constant regurgitation of food, discharge of foul-smelling, slimy faeces and excessive drowsiness due to loss of energy. It can be treated with the drug metronidazole, obtainable from a veterinarian, who will also give details of dosage. At the same time, you can administer a multivitamin solution via a syringe and tube. It is a wise precaution to treat all other snakes in the collection with a preventative drug. Ask your veterinarian for advice.

### Bacterial diseases
Bacterial infections occur in a variety of different forms, but all become significantly more serious when hygiene and general maintenance are poor. If treatment is delayed, such infections can become dangerous in the latter stages for both the snake and its keeper. Here, we look at some common bacterial diseases. However, since there are so many, consult a veterinarian in all cases.

**Pseudomonas bacteria**, in their many forms, are the source of numerous snake problems. To a limited extent, they are present in all healthy reptiles and where conditions are good, they do not become a cause for concern. However, if allowed to flourish, they can cause mouthrot or jaw suppuration, especially in water snakes. Symptoms include a cheesy deposit in the jaws, loss of teeth, jaw gaping and loss of appetite. In addition, the skin around the jaws becomes ulcerated and the linings of the gullet, stomach and intestines can become infected. *Pseudomonas* are difficult to treat effectively, as they tend to recur. A veterinarian will be able to administer an intramuscular injection of a suitable antibiotic, but in cases of mouthrot the infected area should also be dabbed with cotton wool soaked in a 2 percent solution of hydrogen peroxide and then sprayed with a 22 percent aqueous solution of sulphamerazine.

Epidermal necrosis is caused by a different group of *Pseudomonas* bacteria and is common in wild-caught snakes that have been in close contact with infected snakes and faeces during transportation. The usual signs are inflamed lumps and abscesses covering the skin. Apart from the normal treatment for *Pseudomonas*, these abscesses will need lancing and treating with a 3 percent hydrogen peroxide solution, or should be brushed with a strong iodophor (a disinfectant effective against a wide range of pathogens - obtainable from pet stores or your veterinarian).

**Pneumonia** is a bacterial lung infection, most likely to occur when conditions in the vivarium are too cool, damp or inadequately ventilated, or a combination of all three. Symptoms include respiratory problems, with irregular breathing, wheezing and mucal fluid discharged from the nose and mouth. If not treated swiftly, death is inevitable, so keep the infected animal dry and warm and administer a multivitamin solution by mouth, using a syringe without the needle. A veterinarian may prescribe a course of orally administered antibiotics or a series of intramuscular antibiotic injections, depending on the extent of the infection.

**Salmonellosis** is not as common in snakes as in other herptiles.The *Salmonella* bacteria are present in various forms, and harmless quantities occur in most foods, but complications may arise when a snake's resistance is low, following illness, stress,

poor maintenance or eating contaminated food. Severe infections lead to the discharge of green, slimy excreta, regurgitation of undigested food and mucal voiding. The condition can be difficult to treat effectively in its latter stages, but oral antibiotics may help. There is also a risk that the keeper may become infected by certain *Salmonella* bacteria, so always offer fresh food, maintain good hygiene and wash your hands after handling any snakes.

### Nutritional disorders
Unlike other herptiles, snakes do not need a varied diet, because most of them derive all their requirements from one food source. However, nutritional imbalances can occur from time to time, usually in juveniles or in specimens given a poor diet or in those species that feed on insects and other invertebrates. Today, so many multivitamin and mineral preparations are available that these imbalances are easily checked. Simply dust all food items with a multivitamin powder or add a few drops of vitamin liquid to the drinking water.

### Other disorders
Some of the most commonly encountered problems in snakes are the result of incorrect management, such as a poorly designed or inadequately equipped and badly sited vivarium, or overhandling of a specimen.

Stress is one such problem. It can be brought about by any of the factors described above, and can result in a lowering of the

Below: *The blood and pus in this reticulated python's mouth are a sign of mouthrot. Feeding unclean food could be one cause.*

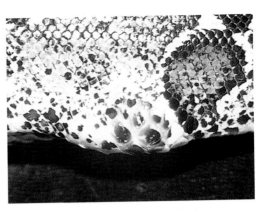

Left: *A prolapsed cloaca in a corn snake. This is commonly seen in egg-laying females and may indicate a dietary deficiency.*

Above: *An abscess in a boa constrictor. Treat the condition swiftly, otherwise it may require an unsightly lancing operation.*

snake's resistance, so that it loses its appetite and succumbs easily to a disease or infection. Egg retention or premature deposition of eggs by females are also possible consequences of stress, as is the sudden death of a seemingly healthy snake. There is no medical cure for stress; the best course of action is to examine how you are maintaining your snakes and ask yourself whether you are disturbing them unnecessarily by constant handling.

## Skin sloughing difficulties

Sometimes, a snake may be unable to shed its outer skin in one complete strip. Instead, it remains attached to the new skin in dried wrinkled shreds or large hardened patches. This can be both irritating and exhausting for the snake, and causes severe stress. Always include a rock with a coarse surface in the vivarium to give the snake a better grip, enabling it to remove its old skin in one piece. If the problem persists, place the snake in a container with some moistened sphagnum moss and a rock. Removing the skin with blunt tweezers is a last resort, since the procedure can prove highly traumatic.

## Wounds

Newly acquired and juvenile snakes often have the annoying habit of butting their heads on the glass front of the vivarium. This can result in a sore, descaled snout that could become infected. Dress the wound with an iodophor or antibiotic solution. This activity usually subsides in time, but in the early stages it may help to provide more hiding places and subdued light outside the vivarium. Always make sure there are no sharp edges in the vivarium.

Occasionally, one snake may be bullied by another and become injured, or it may be necessary to remove an abscess. In either case, dress any injuries with an iodophor or sulphanilamide (available from a veterinarian) and isolate the snake until the wounds heal.

Below: Elaphe quatuorlineata *in the process of sloughing. Opaque eyes are a sign* that the snake is about to shed its skin. Raise humidity levels and provide coarse rocks.

# Snakes
# Species
## Section

The majority of snakes will 'survive' in a relatively small vivarium because of their adaptable body shape. However, in less than spacious surroundings, only the miniature species, such as worm snakes, (*Typhlops*) will breed. Realistically, snakes require a large vivarium if they are to thrive and, although there are no fixed rules regarding ideal vivarium sizes, the following is a useful guide. For every 30cm(12in) of snake, allow 14,750-16,390cm$^3$ (900-1,000in$^3$) of space, i.e. two 60cm(24in) snakes will require a vivarium measuring 60x30x38cm(24x12x15in). If you have limited space, it may be possible to keep different species of snake together in the same vivarium. Compatibility depends upon the species concerned, whether they require the same climatic conditions, their temperament and diet and ultimate size. It is possible to lay down the following guidelines.

**1** Miniature species, e.g. worm snakes (*Typhlops*) cannot be kept together with any other snake family.

**2** Small species, e.g. Dekay's snake (*Storeria*), green snakes (*Opheodrys*) and small garter snakes (*Thamnophis*) can be housed with each other, but not with larger snake species.

**3** Medium-sized species, e.g. water snakes (*Natrix, Nerodia* and *Thamnophis*), whipsnakes and racers (*Coluber*) and ratsnakes (*Elaphe*) can be kept together only if they are of similar size. Bear in mind their country of origin and remember that some species may eat other snakes if they become very hungry.

**4** Large species, e.g. boids and pythons, should only be housed with other members of their own species and then only if there is sufficient space and the snakes are the same size.

**5** Cannibalistic species, e.g. king and milksnakes (*Lampropeltis*) should never be kept with other snake species or even with other members of their own species. House each specimen in its own vivarium, at least until breeding is to be attempted.

On the following pages, you will find details of the most widely available and adaptable species, as well as examples of the most attractive and desirable members of each family.

Below: *Cook's tree boa,* Corallus enhydris cookii, *is very popular, but there are too few captive-bred specimens to meet the growing demand.*

## FAMILY: ANILIIDAE
Pipe snakes

**Distribution:** Southeast Asia and northern South America.
**Length:** Up to 80cm(32in).
**Sex differences:** Use a probe to determine sex.
**Diet:** Because of their rigid jaw, these snakes can only take eels, amphibians, lizards and small rodents, which they constrict or crush with their jaws.
**Ideal conditions:** Subterranean vivarium, with a warm, humid atmosphere, heated to about 28°C(82°F).
**Hibernation:** No.
**In captivity:** Attractive and very easy to keep, given the correct conditions.

These burrowing snakes are only occasionally offered for sale, as their secretive habits make them difficult to find. Their striking appearance and coloration has led to an increase in demand, but little is known about their breeding habits. It seems that the females are ovoviviparous, giving birth to 3-10 relatively large young at any time of year.

The most desirable and easy to keep species are the red and black blotched pipe snake, *Cylindrophis maculatus*, from Sri Lanka and its Venezuelan counterpart the South American pipe snake, *Anilius scytale*.

Below: Anilius scytale. *If attacked, the pipe snake flips over and flattens itself on the ground. By waving its tail, it attracts the predator's attention away from the head.*

## FAMILY: TYPHLOPIDAE
Blind snakes

**Distribution:** Throughout the warmer and tropical zones, including South Africa and southern Australia.

**Length:** Blind snakes range in size from Reuter's blind snake, *Typhlops reuteri*, which rarely attains more than 10cm(4in), to Peter's blind snake, *T. dinga*, at 75cm(30in).

**Sex differences:** In some species, females have large scent glands near the base of the tail, otherwise use a probe to determine sex.

**Diet:** Ants, termites, small crickets, earthworms and slugs.

**Ideal conditions:** Small containers filled with a moist, loose burrowing substrate. Maintain high humidity and a temperature of 24-29°C(75-84°F).

**Hibernation:** No, except for the European worm snake, *Typhlops vermicularis*, and North American species, which should be kept at 8-12°C(14-20°F) lower than normal for a few months before the breeding season.

Above: *The harmless Asian blind snake, Typhlops braminus, closely resembles an earthworm. It has poor eyesight and similar burrowing habits. Easy to keep and feed.*

**In captivity:** North American and European species are easy to maintain, but it is vital to control the temperature more carefully for tropical specimens.

Worm snakes are ideal for beginners, as they require little attention compared to terrestrial or arboreal snakes and are apparently easy to breed. The common Asian species, *Typhlops braminus*, is encountered quite regularly but, unfortunately, other species are rarely offered in the hobbyist market.

Most species are oviparous, depositing 3-12 eggs, which hatch out after a short incubation of 45-60 days at 25°C(77°F). A few species are viviparous and some populations consist entirely of females. In all cases, the young are minuscule and may be difficult to feed in the early stages.

Left: *With acute vision and surprising stealth and agility,* Dryophis nasutus *is the perfect hunter of the trees.*

Right: *Merging in with the surrounding greenery,* Oxybelis *suddenly darts out on its unsuspecting prey.*

Right: *The beautifully marked flying snake,* Chrysopelea ornata, *spends all its time in the trees and bushes.*

Below: *The pointed snout and narrow body of* Oxybelis aeneus *are perfectly suited to its arboreal existence.*

## FAMILY: COLUBRIDAE

This is the largest of all the snake and reptile families, consisting of over 1,200 species. It contains some of the easiest, most interesting and beautiful snakes, many of which are available to the snake-keeper. A few species are slightly venomous, but because their small fangs are located towards the back of the jaw and their venom is weak, they are suitable captive subjects. However, the beautiful mangrove snake, *Boiga dendrophila*, or the boomslang, *Dispholidus typus*, are dangerous, as their venom is particularly potent and you must hold a Dangerous Animals licence to keep these species. As the family is so large and the snakes vary considerably in their appearance and requirements, we shall examine each subfamily individually.

## SUBFAMILY: Boiginae

This subfamily consists mainly of back-fanged venomous snakes, including the deadly boomslang and mangrove snakes. It is safer to wear leather gloves when handling

these snakes, even though in most species the venom is ejected in very small quantities. Here we consider the arboreal species, such as the flying snake, *Chrysopelea ornata*, and the terrestrial species, such as the Montpellier snake, *Malpolon monspessulanus*.

## Flying snake
*Chrysopelea ornata*

**Distribution:** Southeast Asia, India and Sri Lanka.
**Length:** 75-85cm(30-33.5in).
**Sex differences:** Males have a longer tail.
**Diet:** Not at all fussy; eats insects and any small animals.
**Ideal conditions:** A tall planted vivarium, kept warm and humid, at a temperature of 25-29℃(77-84℉).
**Hibernation:** No.
**In captivity:** Relatively easy, but as with all Southeast Asian species, an initial quarantine period is essential. The beautiful flying snake has iridescent colouring and derives its name from its apparent ability to fly. In fact, we now know that it can only glide from one elevated position to another by spreading out its ribs to flatten the body.

**Other species of interest:** Other arboreal species frequently offered for sale include the Asian long nosed tree snake, *Dryophis nasutus*, and various Latin American vine snakes, usually *Oxybelis acuminatus* and *O. fulgidus*. All are diurnal, have acute vision and generally make very interesting captives. Information on their breeding requirements is limited, but they are mainly oviparous, laying 4-12 small eggs.

Left: *Like most other European snakes, the robust* Malpolon monspessulanus *is only occasionally available to hobbyists.*

Right: *Kingsnakes are popular and easy to keep.* Lampropeltis g. holbrooki *is a very attractive subspecies.*

Below right: *The stunning Sonoran mountain kingsnake,* Lampropeltis pyromelana, *breeds regularly in captivity.*

## Montpellier snake
*Malpolon monspessulanus*

**Distribution:** Mediterranean region, North Africa and Eurasia.
**Length:** 130-195cm(51-77in).
**Sex differences:** Males have a longer tail than females.
**Diet:** Lizards, birds and rodents.
**Ideal conditions:** A spacious semi-desert vivarium with plenty of hiding places and a temperature of around 25ºC(77ºF).
**Hibernation:** No, but keep at 18ºC(64ºF) for a few months before breeding.
**In captivity:** A large, formidable snake that can be temperamental and is thus suitable for the more experienced snake-keeper.

**Other species of interest:** Two other European species occasionally become available. One is the European catsnake, *Telescopus fallax*, a small species that rarely attains 65cm(25.5in). Unfortunately, it feeds exclusively on lizards. The false smooth snake, *Macroprotodon cucullatus*, is an equally small lizard-eater and is not recommended. All species are egg layers, depositing relatively small clutches of 4-12 eggs.

SUBFAMILY: Colubrinae
Typical colubrids
This is the main subfamily within the

Colubridae. Most of the attractive and popular species sold by dealers are captive-bred, although some of the recently discovered Russian and other Asian ratsnakes are wild-caught. The subfamily can be considered under four principal headings: the king and milksnakes, the ratsnakes, the whipsnakes and racers and, in the final group, other popular colubrids.

## Common kingsnake
*Lampropeltis getulus*

**Distribution:** Central and southern USA, Mexico.
**Length:** 120-220cm(48-86in).
**Sex differences:** Males have longer tails.
**Diet:** Small mammals and birds, but also frogs, lizards and other snakes.
**Ideal conditions**: A dry desert vivarium with a hot-spot, heated to a temperature of 28ºC(82ºF).
**Hibernation:** No, but reduce temperature to 19ºC(66ºF) several months before breeding.
**In captivity:** Easy to maintain, but house individually to avoid the risk of cannibalism. With their bright colours arranged in two, three or sometimes four bands, the tricoloured kingsnakes and milksnakes are the most stunningly attractive of all serpents. Although very expensive to obtain, even as

hatchlings, they will thrive given good care and eventually reward their keeper by breeding regularly. Many species and subspecies are currently available, but many are incorrectly named because they resemble one another so closely. All kingsnakes share the same requirements with regard to temperatures, feeding and hibernation, but vivarium size may vary considerably.

**Other species of interest:** The most popular subspecies is undoubtedly the Californian kingsnake, *L.g. californiae*, which comes in two forms: the common black or brown-and-white banded, and the rarer black with three longitudinal white stripes. Other highly attractive subspecies include the Florida king, *L.g. floridana*, the speckled king, *L.g. holbrooki* and the Mexican king, *L.g. nigritus*. Occasionally, mutant albino forms of all the above are available.

## Popular kingsnakes and milksnakes

| Species | Size | Difficulty in captivity |
|---|---|---|
| Californian mountain kingsnake (*L. zonata*) | 75-100cm(30-39in) | Can be a fussy feeder, preferring lizards. |
| Sonoran mountain kingsnake (*L. pyromelana*) | 65-90cm(25-36in) | Easy to keep and breed. |
| Mexican kingsnake (*L. mexicana*) | 75-100cm(30-39in) | Easy to keep and breed. |
| Milksnake (*L. triangulum*) | The nominate subspecies *L.t. triangulum* is rather dull, but other subspecies are some of the most beautiful snakes and extremely popular in captivity. | |
| Mexican milksnake (*L.t. annulata*) | 60-80cm(24-32in) | A small, easily kept subspecies. |
| Pueblan milksnake (*L.t. campbelli*) | 75-95cm(30-37in) | Easy to keep and breed. |
| Honduran milksnake (*L.t. hondurensis*) | 150-230cm(60-90in) | Sometimes finicky, and wild specimens may not settle at all. Still a very popular snake. |
| Sinaloan milksnake (*L.t. sinaloae*) | 70-100cm(27-39in) | Possibly the most beautiful serpent. Rare, but easy to keep and breed. |

Above right and right: *These two photographs show the two forms of the Californian kingsnake,* L.g. californiae. *The banded form (top) is the most often seen. The striped variety occurs only in the vicinity of San Diego.*

Left: *The harmless Californian mountain king,* Lampropeltis zonata *is sometimes mistaken for its poisonous cousin,* Micruroides euryxanthus, *the Sonoran coral snake.*

Breeding usually takes place once or, very rarely, twice a year between March and July. Up to 20 eggs are laid and these require an incubation period of around 60 days at 28⁰C(82⁰F). The young measure 30cm(12in) and are able to feed on newborn mice. As all these snakes have cannibalistic tendencies, it is best to raise them individually in small plastic containers from the outset.

### Corn or red ratsnake
*Elaphe guttata*

**Distribution:** Eastern USA.
**Length:** Up to 180cm(72in).
**Sex differences:** Males have a longer tail.
**Diet:** Small rodents and birds.
**Ideal conditions:** To accommodate a pair, provide a 90cm(36in) dry woodland vivarium with plenty of branches. Maintain the temperature at around 28⁰C(82⁰F).
**Hibernation:** No, but reduce temperature to around 19⁰C(66⁰F) for several months.
**In captivity:** An easy, placid species, ideal for the beginner, but predominantly nocturnal.

Above: Elaphe guttata, *the corn snake, easily bred and kept in captivity.*

Below: *One of the most sought after serpents, the leopard snake,* Elaphe situla.

140

Ratsnakes are very popular, mainly because they are unfussy feeders and, given the correct conditions, will breed regularly. All ratsnakes breed up to three times a year. Clutch sizes may be as small as three eggs in *E. situla*, but up to 25 eggs in *E. guttata*. They require between 45 and 90 days incubation and the hatchlings will feed on pinkie mice and grow swiftly.

**Other species of interest:** Other American ratsnakes have more arboreal habits and are suited to a tall or woodland vivarium. The black rat, *E. obsoleta obsoleta*, is a common species that regularly grows to 200cm(79in) or more, and is mainly diurnal. It has a number of appealing and distinct subspecies, including the yellow, *E.o. quadrivittata*, the grey, *E.o. spiloides*, and the Texas ratsnake, *E.o. lindheimeri*. One of the prettiest is the smaller Trans-Pecos ratsnake, *E. subocularis*, from southern USA and northern Mexico.

In the last few years, Asian ratsnakes have appeared with greater regularity on dealers' lists. The Russian ratsnake, *E. schrenki*, and Dione's ratsnake, *E. dione*, make excellent small captives, but larger types, such as the copperhead ratsnake, *E. radiata*, and the hundred flower ratsnake, *E. moellendorfii*, may attain 213-300cm(84-120in) and therefore need extensive quarters, especially to breed. Other Asian ratsnakes called Dhamans, *Ptyas mucosus* and *P. kovcos*, grow even larger, but have a nasty disposition, so only the experienced snake-keeper should attempt to keep them.

European ratsnakes are not as frequently available as their American and Asian counterparts, as they are totally protected in the wild. Captive-bred specimens are occasionally listed, including the Aesculapian snake, *E. longissima*, the four-lined snake, *E. quatuorlineata*, and the most beautiful serpent of the whole genus, the leopard snake, *E. situla*. Unfortunately, it has a tendency to refuse food and, as it is quite scarce, you should only attempt to keep it when you have more experience.

Below: *Opaque eyes and dull skin colour on the head of* Elaphe obsoleta lindheimeri *are signs that it will soon shed its old skin.*

141

**Racer**
*Coluber constrictor*

**Distribution:** Throughout North America.
**Length:** 80-240cm(32-95in).
**Sex differences**: Males are smaller and have a longer tail.
**Diet:** Large insects, frogs and small mammals.
**Ideal conditions:** Spacious dry quarters, with plenty of branches and a hot-spot. Temperatures around 25°C(77°F).
**Hibernation:** Yes.
**In captivity:** A nervous and sometimes aggressive species, but easy to maintain.

All whipsnakes and racers are highly active, semi-arboreal and sun-loving. They dislike being handled and tend to be rather aggressive. They are ideally suited to a large vivarium or, better still, a greenhouse enclosure, rather than a small cage. Reports of breeding success are rare at present, probably because little is known about their habits and requirements, although exposure to ultraviolet light is undoubtedly beneficial. Small clutches of 5-20 eggs are laid in early summer, and these should be incubated in the normal way. Hatchlings tend to be rather small and weak and should be fed on locust-hoppers or crickets, liberally dusted with multivitamin powder.

**Other species of interest:** The racer is more likely to appear on a dealer's list than other species of whipsnake and racer, but occasionally there may be more choice. American whips and racers are very attractive, especially the smaller striped whipsnake, *Masticophis taeniatus*, and the pencil-like coachwhip, *M. flagellum*, which can grow to over 3.4m(11ft).

All the European species available today are captive-bred, and are therefore in rather short supply compared to demand. The western whipsnake, *Coluber viridiflavus*, the horseshoe whip, *C. hippocrepis*, the beautiful Dahl's, *C. najadum*, and the robust Balkan, *C. gemonensis*, are some of the choicest and most desirable examples.

**Other popular colubrids**
Many other colubrids appear regularly in dealers' lists and are worth mentioning here. Serpents of the North American genus *Pituophis*, namely the bull, gopher and pine snakes, are very robust and grow to 244cm(96in), and are only suitable for hobbyists with adequate space. They require large, dry enclosures and will breed regularly if the temperature is first lowered to around 12°C(54°F). Between 4 and 16 eggs are deposited and the large hatchlings appear after about 60 days. They are able to take small mice and mature within two years.

The Florida indigo snake, *Drymarchon corais couperi*, grows even larger than *Pituophis* snakes, reaching up to 300cm(120in) in length, but requires very similar conditions. Take care, particularly when cooling down males before breeding, as they can become rather irascible and their bite is quite capable of causing lacerations.

American green snakes, the rough *Opheodrys aestivus* and the smooth *O. vernalis*, are ideal for beginners. They rarely attain 65cm(25.5in), have a completely insectivorous diet, consisting of grubs, crickets and locusts, and require small cages and similar care to garter snakes, *Thamnophis* spp. If hibernated at 10°C(50°F) for several months, they may mate and deposit 3-8 tiny, fragile eggs. These have a short incubation period of around 45 days and the minuscule hatchlings will need constant attention for the first four months. They are best reared individually in plastic boxes.

The house snakes, *Lamprophis* spp., are just one of a multitude of interesting African genera, including egg-eating snakes, *Dasypeltis* spp. and the filesnakes, *Mehelya* spp., which have not as yet become very popular in the Northern Hemisphere, possibly because they are difficult to obtain in sufficient numbers. The brown house snake, *L. fuliginosa*, is a relatively small species, growing to 107cm(42in), that seems

to thrive in captivity. It breeds throughout the year if cooled down periodically for six weeks to around 15°C(59°F), normally producing small clutches of 3-12 eggs. Hatchlings feed on pinkies, eventually progressing to adult mice or small rats.

SUBFAMILY: Natricinae
Water snakes
This important subfamily includes some of the best-known and most popular snakes, especially for the beginner. All species are small to medium in size, inhabiting

Below: *The gopher snake,* Pituophis melanoleucus, *is a robust, diurnal constrictor that adapts well to life in captivity.*

Left: *Take extreme care whenever you handle whipsnakes or racers, such as this* Coluber constrictor mormon. *They are very agile, and thrash about and bite if not restrained.*

Right: *Secretive and well camouflaged, the western whipsnake,* Coluber viridiflavus, *hunts for lizards and fledglings in the trees.*

meadows, woodland and forest, but never found too far away from water. There are two distinct types of Natricinae, the grass snakes and the garter snakes.

## Grass snake
*Natrix natrix*

**Distribution:** Europe, Asia east to Lake Baikal, and North Africa.
**Length:** 90-200cm(36-79in).
**Sex differences:** Females are usually larger than males.
**Diet:** Earthworms, fish, amphibians and smaller rodents.
**Ideal conditions:** Either a spacious, semi-aquatic set-up, which must include a dry area, or a hygienic vivarium with a basking area. Do not allow the vivarium to become too warm - around 22-27°C(72-81°F) is sufficient.
**Hibernation:** Yes.
**In captivity:** Usually very easy to maintain, but some individuals may not adapt to captivity. Remember that all water snakes must have access to a fairly large dry area, as constant contact with water will lead to sores.

All grass snakes are oviparous, producing 12-50 eggs in early summer. The young hatch out after 45-50 days and feed on earthworms and small fish.

**Other species of interest:** Many attractive subspecies of grass snake are available from time to time. Other recommended Eurasian species include the smaller viperine snake, *N. maura* - which, as its common name suggests, resembles a viper - and the dice snake, *N. tessellata*. The checkered keelback, or Asian water snake, *Natrix piscator*, is very common on reptile dealers' lists and makes a good vivarium subject. It grows to around 120cm(48in) and may lay up to 80 eggs.

The North American water snakes were once classified in the genus *Natrix*, but are now in a separate genus called *Nerodia*. There are several attractive species and all require similar conditions and maintenance to the European and Asian water snakes. All known species are livebearing, depositing up to 85 minute young in midsummer. Unfortunately, they dislike earthworms, so offer the young snakes thin strips of fish. The northern water snake, *Nerodia sipedon*, and

| Popular garter and ribbon snakes | | |
|---|---|---|
| Species | Size | Distribution |
| Western garter (*T. elegans*) | 65-95cm(25-37in) | Western USA |
| Mexican garter (*T. eques*) | up to 130cm(50in) | Southern USA and Mexico |
| Northwestern garter (*T. ordinoides*) | 50-80cm(20-32in) | Northwest USA |
| Plains garter (*T. radix*) | 60-85cm(24-33in) | North and Central USA |
| Western ribbon (*T. proximus*) | 50-130cm(20-50in) | Central and Southeast USA |
| Common garter (*T. sirtalis*) | many attractive subspecies are available. The following are the most popular:- | |
| Red-spotted garter (*T.s. concinnus*) | 65-90cm(25-36in) | West coast USA |
| Valley garter (*T.s. fitchii*) | 55-75cm(21-30in) | Western USA |
| Red-sided garter (*T.s. parietalis*) | 60-85cm(24-33in) | Throughout USA |
| Eastern garter (*T.s. sirtalis*) | up to 120cm(48in) | Eastern USA |

Above: Thamnophis sirtalis parietalis *can grow to 85cm(33in) in suitable conditions.*

Left: Natrix natrix, *the grass snake, is one of the most popular snakes for the beginner.*

Below: *The attractive checkered garter snake,* Thamnophis marcianus, *from southern USA and Mexico, is just one of the easily kept species and subspecies of garter snake available.*

the large red-billed water snake, *Nerodia erythrogaster*, are the species most commonly offered within the hobby.

## Blue-sided garter snake
*Thamnophis sirtalis similis*

**Distribution:** Eastern USA, especially Florida.
**Length:** 80-110cm(32-43in).
**Sex differences:** Males have a longer tail, but females are heavier and larger overall.
**Diet:** Earthworms, fish, amphibians and, occasionally, small rodents.
**Ideal conditions:** A woodland or hygienic vivarium, with a large water bowl. Maintain the temperature at between 22-25⁰C(72-77⁰F).
**Hibernation:** Yes, at around 9⁰C(48⁰F) for a few months.
**In captivity:** Probably the easiest snake to maintain, although breeding can be somewhat erratic.

Breeding is somewhat irregular, but more likely to occur immediately after a period of hibernation. It is a good idea to keep several males and females in one large vivarium to ensure at least one successful mating. All garter snakes are ovoviviparous and deposit 10-75 young on damp sphagnum moss after a gestation period of about 50 days. Initially, the young are very small and fragile and best raised individually in small plastic containers. They will accept strips of fish or, occasionally, strips of raw meat and all but the ribbon snakes will eat earthworms. Maturity is attained within a year if the snakes are well maintained.

**Other species of interest:** Many species and subspecies of garter and ribbon snakes are available, all varying in size and coloration. The most popular species and subspecies are

145

listed in the panel and all require similar care and conditions, except perhaps for southern species, which benefit from a few extra degrees of warmth during periods of activity and winter rest.

SUBFAMILY: Xenodontinae
American hognose snakes
This small subfamily also includes a few back-fanged venomous species. Although very familiar to Americans, the hognose snakes, *Heterodon* spp., are only now becoming popular in European countries, mainly because captive-breeding now occurs on a regular basis.

**Western hognose**
*Heterodon nasicus*

**Distribution:** Southern Canada to northern Mexico.
**Length:** 45-90cm(18-36in).
**Sex differences:** Usually determined by probing.
**Diet:** Fish, amphibians, small lizards and small rodents.
**Ideal conditions:** A small dry vivarium, heated to a temperature of about 24-27°C (75-81°F).
**Hibernation:** Yes.
**In captivity:** If they can be persuaded to eat rodents, these snakes are very undemanding vivarium subjects. Can breed each year.

Above: *The Western hognose,* Heterodon nasicus, *is widely available. Captive-bred juveniles make good vivarium subjects .*

Above: Heterodon platyrhinos, *the eastern hognose. When molested, it displays an amazing reaction, flipping onto its back and feigning death by opening its mouth.*

*H. nasicus* varies in both coloration and appearance, and there are a number of subspecies. The usual colour is sandy brown, but bright orange individuals have been known. The name 'hognose' refers to the hornlike adornment that is thought to be used for burrowing.

Hognoses mate immediately after hibernation and deposit 8-20 eggs in late spring after a short gestation. The relatively robust young snakes hatch out after 50 days and feed on small fish or frogs. They grow quickly and can mature within two years.

**Other species of interest:** Another species occasionally available is the larger eastern American hognose, *Heterodon platyrhinos.* Unfortunately, this species seems to eat nothing but toads, although it is otherwise easy to maintain.

## FAMILY: BOIDAE

The snakes in this family are some of the most popular, both with serious hobbyists and among people who have only a casual interest in snakes. Many are kept singly as pets, which is a pity, since smaller species can easily be persuaded to breed. All boids are constrictors, famed for their ability to swallow food far larger than the size of the mouth opening would suggest. However, here we are concerned only with those species that attain a relatively short length. Potentially massive types, such as the anaconda, *Eunectes murinus*, the reticulated python, *Python reticulatus*, the African rock python, *P. sebae*, and the Indian python, *P. molurus*, have been omitted, not only because they would eventually require huge enclosures, but also because of the risk involved in being attacked during cleaning out or handling. The Boidae family is divided into four subfamilies, but here we shall look only at the two major subfamilies, the boas and the pythons.

Below: *An adult* Boa constrictor. *Handle these impressive snakes with the utmost care and respect at all times, as they have immensely powerful coils and sharp teeth.*

SUBFAMILY: Boinae
Boas

Boas are some of the most popular serpents and in recent years, breeding results have improved dramatically, leading to the availability of healthy, captive-bred juveniles that adapt far better to an artificial environment. They differ from their close relatives, the pythons, in that they give birth to live young. In all boas, a small (and unnatural) drop in temperature usually induces reproductive behaviour. After a successful union, a long gestation period of 100-300 days follows, after which the tree boa's will produce just one offspring, whereas boa constrictors may have up to 50. The young are large and easily able to deal with pinkies and lizards, but they are best raised in individual containers for the first six to eight months. They can attain maturity within 18 to 24 months.

### Boa constrictor
*Boa constrictor*

**Distribution:** Central and South America.
**Length:** 200-300cm(79-120in).
**Sex differences:** Males have larger spurs near the cloaca.

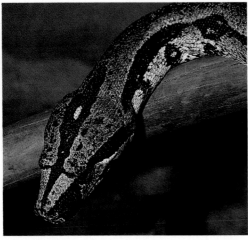

Left: Boa constrictor imperator. *Captive-bred specimens of this fine red-tailed snake are much sought after by keen collectors.*

Above: *There are several subspecies of rainbow boa. The Brazilian subspecies,* Epicrates c. cenchria, *is the most desirable.*

### Rainbow boa
*Epicrates cenchria*

**Distribution:** Central and South America.
**Length:** 160-200cm(63-79in).
**Sex differences:** Males have larger spurs near the cloaca.
**Diet:** Small rodents and birds.
**Ideal conditions:** A large and fairly tall cage, with several robust branches. Maintain temperature at around 29°C(84°F).
**Hibernation:** No, but provide a 'cool' period of 18°C(64°F) for four to six weeks before breeding.
In captivity: A beautiful and desirable species that settles down well in captivity.

**Diet:** Rodents, rabbits and birds.
**Ideal conditions:** A large, lofty vivarium, with several strong branches. Maintain the temperature at around 29°C(84°F).
**Hibernation:** No, but a 'cool' period at 18°C(64°F) for four to six weeks before breeding is beneficial.
**In captivity:** Sometimes temperamental, but captive-bred specimens make excellent, long-lived vivarium subjects.

**Other species of interest:** A more attractive, but more expensive, red-tailed form, *Boa constrictor imperator*, is frequently available and requires similar care.

**Other species of interest:** Other species of *Epicrates* occasionally become available, including the large Haitian boa, *E. striatus*, which is far more temperamental than its rainbow cousin, and the pretty Cuban boa, *E. angulifer*, which can be highly aggressive and should only be tackled by more experienced hobbyists.

## Cook's tree boa
*Corallus enhydris cookii*

**Distribution:** South America.
**Length:** 110-150cm(43-59in).
**Sex differences:** Usually determined by probing.
**Diet:** Small rodents and birds.
**Ideal conditions:** A tall vivarium, with plenty of strong branches, a warm, humid atmosphere and a temperature of about 27°C(81°F).
**Hibernation**: No, but cool down to 20°C(68°F) for eight weeks before breeding.
**In captivity:** A smaller, highly desirable arboreal boa that adapts well to captivity but dislikes disturbance.

**Other species of interest:** At present, tree boas are not frequently bred in captivity, which is a great pity since there is an enormous demand for them. The more familiar emerald tree boa, *Corallus canina*, is a beautiful species but, unfortunately, it can prove difficult to maintain and is more temperamental than Cook's tree boa. To begin with, the newborn of all species will only accept a diet of lizards.

Above: *Cook's tree boa,* Corallus enhydris cookii, *needs a spacious vivarium with plenty of lofty branches.*

Below: *An adult* Corallus canina. *Initially, the young are brick red, becoming emerald green some two years later.*

**Turkish sand boa**
*Eryx jaculus*

**Distribution:** Southeastern Europe, western Asia and North Africa.

**Length:** Rarely more than 80cm(32in).

**Sex differences:** Males have larger spurs near the cloaca.

**Diet:** Earthworms, nestling birds and small rodents.

**Ideal conditions:** A small, dry, desert vivarium, with a temperature in the region of 25-28°C(77-82°F).

**Hibernation:** No, but cool down to around 14°C(57°F) for eight weeks in the winter.

**In captivity:** An easy but very timid species.

**Other species of interest:** The genus *Eryx* consists of 11 species and many subspecies, all of which rarely grow more than 90cm(36in) in length. They are not as popular as other boas, but are relatively easy to breed if a 'cool' period is introduced. The rough-scaled sand boa, *Eryx conicus*, and the Indian sand boa, *E. johni*, are occasionally available to the hobbyist.

Above: *Like all sand boas, the eastern European species,* Eryx jaculus, *is equipped with a blunt snout for burrowing.*

Left: *Sand boas, such as* Eryx johni, *have short bodies with smooth scales. They can glide easily through desert sands.*

## FAMILY: PYTHONINAE
Pythons

Pythons are oviparous serpents and, in certain species, the female wraps her body around the eggs and incubates them until they start to hatch. This behaviour is unusual in snakes. Most pythons are sizeable animals, best left to experienced hobbyists or zoological collections. However, there are a limited number of attractive species suitable for the amateur hobbyist.

### Green tree python
*Chondropython viridis*

**Distribution:** Papua New Guinea and the northern tip of Australia.
**Length:** 100-140cm(39-55in).
**Sex differences:** Usually determined by probing.
**Diet:** Mainly birds but also rodents.
**Ideal conditions:** A tall vivarium with plenty of branches. Keep it warm and humid with a temperature of about 29$^0$C(84$^0$F).
**Hibernation:** No, but cool down to 19$^0$C(66$^0$F) for four to six weeks before breeding.
**In captivity:** Extremely beautiful and desirable but expensive. Much better temperamentally than the almost identical emerald tree boa, *Corallus canina*, but it does grow larger and is harder to breed.

### Ball or royal python
*Python regius*

**Distribution:** Western Africa.
**Length:** 90-140cm(36-55in).
**Sex differences:** Usually determined by probing.
**Diet:** Small rodents.
**Ideal conditions:** A small 'hygienic' vivarium with a temperature range of 25-29$^0$C(77-84$^0$F).
**Hibernation:** No.
**In captivity:** A docile species, but try to obtain captive-bred specimens, as wild-caught ones very rarely adapt to captivity.

Like boas, pythons benefit from a 'cooling down' period before breeding. After mating, the royal python produces a clutch of 2-10 eggs, whereas the green tree python lays as many as 30. Either remove them or allow the female to incubate them herself. The incubation period lasts up to 75 days at 29$^0$C(84$^0$F). Remove the small pythons to individual containers and feed them on pinkies or day-old chicks.

**Other species of interest:** The Australasian childrens' python, *Liasis childreni*, is smaller (120cm/48in) and regularly offered by dealers. It is ideal for hobbyists wanting to keep a python for the first time.

Below: Python regius *is sometimes called the ball python, because of its habit of rolling itself into a tight ball when molested.*

Right: *An adult* Chondropython viridis *lies in wait for a bird. The young snakes are brick red, later turning yellow.*

# LIZARDS SPECIES INDEX

# SNAKES SPECIES INDEX

Page numbers in **bold** indicate major references including accompanying photographs. Page numbers in *italics* indicate captions to other illustrations. Less important text entries are shown in normal type.

# PICTURE CREDITS

# ACKNOWLEDGEMENTS

## Publisher's acknowledgements

The publishers wish to thank Thorne Electrim, Hants, Ultratherm Heating Ltd., Scotland and The Vivarium, London for their help.

## Author's acknowledgements

Jane Dyke at Purrfect Pets, Phil Canty and Janet Staniszweski.

# FURTHER READING

Alderton, D. *A Petkeeper's Guide to Reptiles and Amphibians* Salamander Books, 1986

Arnold, E.A./Burton, J.A./Ovenden, D.W. *A Field Guide to the Reptiles and Amphibians of Europe* Collins, 1978

Burton, M. *The Encyclopedia of Reptiles and Amphibians* Octopus, 1975

Cooper, J.E. & Jackson, O.F. *Diseases of the Reptilia* Academic Press, 1981

Gans, C. *Reptiles of the World* Ridge Press/Bantam Books, 1975

Mattison, C. *Lizards of the World* Blandford Press, 1989

Mattison, C. *Snakes of the World* Blandford Press, 1986

Mattison, C. *The Care of Reptiles and Amphibians in Captivity* Blandford Press, 1982

Riches, R.J. *Breeding Snakes in Captivity* Arco, 1976

Stebbins, R.C. *A Field Guide to eastern Reptiles and Amphibians* H.M. Co, 1966

Stebbins, R.C. *A Field Guide to western Reptiles and Amphibians* H.M. Co, 1966